T0135959

Strong L^p-Solutions for Fluid-Rigid Body Interaction Problems

Vom Fachbereich Mathematik

der Technischen Universität Darmstadt

zur Erlangung des Grades eines

Doktors der Naturwissenschaften

(Dr. rer. nat.)

genehmigte

Dissertation

von

Dipl.-Math. M.A. Karoline Götze

aus Aachen

Referent:	Prof. Dr. Matthias Hieber
Korreferenten:	Prof. Dr. Dieter Bothe
	Prof. Dr. Yoshihiro Shibata
Tag der Einreichung:	15. Oktober 2009
Tag der mündlichen Prüfung:	22. Dezember 2009

Darmstadt 2010

D 17

Bibliografische Information der Deutschen Nationalbibliothek

Die Deutsche Nationalbibliothek verzeichnet diese Publikation in der
Deutschen Nationalbibliografie; detaillierte bibliografische Daten sind
im Internet über http://dnb.d-nb.de abrufbar.

ISBN 978-3-8325-2599-6

Logos Verlag Berlin GmbH
Comeniushof, Gubener Str. 47,
10243 Berlin
Tel.: +49 (0)30 42 85 10 90
Fax: +49 (0)30 42 85 10 92
INTERNET: http://www.logos-verlag.de

Contents

Introduction

We consider the system of equations which describe the motion of a rigid body in a viscous incompressible fluid. We give a brief introduction to the model and to notation, illustrated by the figure below. The rigid body occupies a bounded domain $\mathcal{B}(t)$ and the fluid fills the exterior domain $\mathcal{D}(t) := \mathbb{R}^3 \backslash \overline{\mathcal{B}(t)}$.

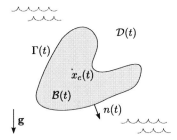

Their interface is denoted by $\Gamma(t)$ and we write

$$\mathbb{R}_+ \times \mathcal{D}(\cdot) := \{(t,x) \in \mathbb{R}^4 : t \in \mathbb{R}_+, x \in \mathcal{D}(t)\}$$

and similarly, $\mathbb{R}_+ \times \Gamma(\cdot)$. The fluid's motion is governed by the Navier-Stokes equations,

$$\begin{cases} v_t + (v \cdot \nabla)v = f + \operatorname{div} \mathbf{T}(v,q) & \text{in } \mathbb{R}_+ \times \mathcal{D}(\cdot), \\ \operatorname{div} v = 0 & \text{in } \mathbb{R}_+ \times \mathcal{D}(\cdot), \\ v = v_\mathcal{B} & \text{on } \mathbb{R}_+ \times \Gamma(\cdot), \\ v(0) = v_0 & \text{in } \mathcal{D}(0), \end{cases} \qquad (1)$$

where v and q denote the velocity and pressure of the fluid and constant density 1 is assumed. The fluid's stress tensor is given by

$$\mathbf{T}(v,q) = \mu_0 \mathcal{E}^{(v)} - q\mathrm{Id},$$

where $\mathcal{E}^{(v)} := \frac{1}{2}(\nabla v + (\nabla v)^T)$ yields the shear rate tensor and the viscosity $\mu_0 > 0$ is constant. The fluid equations are complemented by the following equations on the acceleration of the rigid body,

$$\begin{cases} \mathrm{m}\eta'(t) + \int_{\Gamma(t)} \mathbf{T}(v,q)(t,x)n(t,x)\,\mathrm{d}\sigma = \mathbf{F}(t), \quad t \in \mathbb{R}_+, \\ (J\omega)'(t) + \int_{\Gamma(t)}(x - x_c(t)) \times \mathbf{T}(v,q)(t,x)n(t,x)\,\mathrm{d}\sigma = \mathbf{M}(t), \quad t \in \mathbb{R}_+, \\ \eta(0) = \eta_0, \\ \omega(0) = \omega_0, \end{cases} \quad (2)$$

which contain, in particular, the drag force and torque exerted by the fluid onto the body. Here, m and J are the body's mass and inertia tensor, x_c is the position of its center of gravity and η and ω denote its translational and angular velocity, respectively, so that

$$v_B(t,x) := \eta(t) + \omega(t) \times x$$

is its full velocity. The functions $f, \mathbf{F}$ and $\mathbf{M}$ denote external forces and torques. To model a free fall of the body under the influence of gravitation, we set $f = \mathbf{g}$, $\mathbf{F} = \mathrm{mg}$ and $\mathbf{M} = 0$ for a constant vector $\mathbf{g}$. By $n(t)$ we denote the normal vector at the interface $\Gamma(t)$ which points towards the fluid.

The study of the motions of rigid bodies immersed in a fluid is a classical problem of fluid mechanics. For example, the works of Stokes, [Sto51], and Kirchhoff, [Kir69], mark first seminal contributions to this problem dating back to the mid 19th century. Rigorous mathematical studies on the coupled problem were initiated much later in the works of Weinberger and Sauer, cf. [Wei73] and [Sau72], who investigated the stationary problem. In [Ser87] it was shown by Serre that for every geometry of the rigid body, at least one stationary solution exists.

Starting from these investigations, extensive and pioneering work related to the stationary and instationary problem is due to Galdi and several of his co-authors. Their results concern prescribed, steady, self-propelled and free movements of the rigid body and range from existence theorems of weak and strong solutions to geometric analysis of problems relevant for sedimentation applications in Newtonian and non-Newtonian liquids, cf. e.g. [GHS97], [Gal99], [GV01], [GVP+02], [GS02], [GS05], [Gal06], [Gal07], [GS07]. A first summary and detailed discussion of problems and results can be found in [Gal02].

The existence of global weak solutions for the instationary coupled problem given by (1) and (2) was established by Desjardins and Esteban in [DE99] on bounded domains $\mathcal{D}(t)$ and then by Conca, San Martín and Tucsnak in [CSMT00], cf. also [GLS00]. Later, the weak solution was shown to be unique under smallness assumptions on the data and to exist regardless of possible contacts of body and wall, cf. [Fei03], and [HS99] and [SMST02] for the two-dimensional case. A basic difficulty of the study of weak solutions lies in providing a weak formulation of the problem. Test functions have to suit the fluid part and at the same time respect the space of rigid velocities describing the motion of the body.

Both in the case when weak and when strong solutions for this problem are considered, the methods are strongly related to the existence and regularity theory for the Navier-Stokes equations. The extensive literature on this fundamental problem was developed in particular due to the groundbreaking works of Leray [Ler34] and Hopf [Hop51]. Since then, regular solutions were shown to exist on small time intervals, whereas global solutions are either weak or they require small initial data. The basic problem of global regularity of solutions in three space dimensions remains open and one of the major challenges in mathematics. For an overview of the theory, we refer to the monographs by Temam [Tem79], Galdi [Gal94] and Sohr [Soh01].

In the strong setting, in order to make the classical methods and results from this theory accessible, the first step is to transform (1) to a fixed domain, e.g. $\mathcal{D} := \mathcal{D}(0)$. There are basically two possibilities for the transform. In order to describe the first one, which is linear, we first consider the uncoupled case, where the body velocities ω and η are prescribed. Thus the boundary condition as well as the domain $\mathcal{D}(t)$ in (1) are known for every $t > 0$. Linearly rotating and shifting space back to the original position leads to the rotational Navier-Stokes problem

$$\left\{ \begin{aligned} \partial_t u - \Delta u + \Omega \times u - (u_\mathcal{B} - u \cdot \nabla)u + \nabla p &= f &&\text{in } \mathbb{R}_+ \times \mathcal{D}, \\ \operatorname{div} u &= 0 &&\text{in } \mathbb{R}_+ \times \mathcal{D}, \\ u &= u_\mathcal{B} &&\text{on } \mathbb{R}_+ \times \Gamma, \\ u(0) &= v_0 &&\text{in } \mathcal{D}, \end{aligned} \right. \tag{3}$$

where u, p and $u_\mathcal{B}(t, x) = \xi(t) + \Omega(t) \times x$ denote the transformed fluid and body velocities and pressure and where $\Gamma := \Gamma(0)$. It was shown by Hishida in [His99] that the operator L given by

$$Lu := P_{\mathcal{D},2}(\Delta u + (\Omega \times x \cdot \nabla)u - \Omega \times u),$$

which corresponds to this problem in the case of constant rotational velocity without translation, generates a C_0-semigroup on $L^2_\sigma(\mathcal{D})$ which is not analytic. Here, $P_{\mathcal{D},2}$ denotes the Helmholtz projection on $L^2(\mathcal{D})$. The existence of local mild solution was extended to the spaces $L^q_\sigma(\mathcal{D})$, $1 < q < \infty$, in [GHH06a]. In [FNN07], the essential spectrum of L was shown to consist of equally spaced half lines in the left plane. One of the fundamental difficulties of this approach therefore arises regardless of the coupling, as the transformed system (3) is not parabolic. Additional results on this problem are due to Galdi and Silvestre. In [GS05], it is shown that (3) admits strong L^2-solutions which are even global and converge to a steady state solution if v_0 and Ω are sufficiently small. In [HS09], L^p-L^q estimates on the semigroup generated by L were proved to show a similar result on strong solutions in the L^p-setting.

In [GS02], Galdi and Silvestre used the approach via this transform to show the existence of strong L^2-solutions to the coupled problem combining (1) and (2). They apply a Galerkin-method to the transformed system, where the underlying Hilbert space has to be constructed to respect the body's geometric properties and restrict the movement of the interface to rigid motions.

A second approach is characterized by a non-linear, "local" change of coordinates which only acts in a suitable bounded neighborhood of the obstacle. This idea is motivated by the work of Inoue and Wakimoto on the Navier-Stokes equations on bounded moving domains, cf. [IW77]. Under this transform, the fluid equations are parabolic. A group of authors around Tucsnak, Cumsille and Takahashi used this transform to show the existence of a unique strong L^2-solution to the coupled problem on bounded and unbounded fluid domains in two and three space dimensions, cf. [Tak03], [TT04], [CT06] and [CT08]. In the latter work, not rigid obstacles, but objects which may move in a way to resemble fish-like swimming are considered. Again, in the L^p-setting, there is a result on using this transform on the uncoupled problem. In [DGH09], the existence of a unique strong L^p-solution for the Navier-Stokes equations in the exterior of several obstacles moving with a time-dependent presribed velocity was shown. If the rotational speed is sufficiently smooth, the linearized transformed system can be considered as a type of perturbation of the Stokes problem which is small on small time intervals. Maximal L^p-regularity of this problem can therefore be shown by referring back to maximal regularity of the Stokes problem. The estimates thus obtained allow for solving the non-linear problem via a fixed point argument. Furthermore, it is shown that this strong solution and the mild solution arising from the linear transform coincide in the very weak sense.

In this context, we state the first main result of this thesis. We prove the existence of a unique local strong L^q-solution for the fully coupled problem given by (1) and (2), using a local change of coordinates. A modification of this result for bounded fluid domains is given in Section 6.1. It is a generalization of [Tak03, Theorem 9.1] to the L^p-setting. Note that Takahashi uses a special semigroup approach to solve the transformed linearized problem, which substantially draws on Hilbert space techniques, modified in an involved way to fit the coupling. It is unclear how to transfer this idea to $L^q(\mathcal{D})$, $q \neq 2$, so that we choose a completely different method for showing maximal regularity of the linear system of equations, cf. Section 3. The advantage of this technique is that it can easily be modified to fit the cases of unbounded or bounded fluid domains and the two-dimensional setup, cf. Chapter 6.

Furthermore, it can be extended to a generalization of the equations in (1) and (2) to a class of non-Newtonian fluids. The stress tensor $\mathbf{T}$ is replaced by

$$\mathbf{T}^\mu(v, q) := \mu(|\mathcal{E}^{(v)}|_2^2)\mathcal{E}^{(v)} - q\mathrm{Id},$$

where the viscosity μ is a function in $C^{1,1}(\mathbb{R}_+; \mathbb{R})$ satisfying

$$\mu(s) > 0 \quad \text{and} \quad \mu(s) + 2s\mu'(s) > 0 \quad \text{for all } s \geq 0, \tag{4}$$

and where $|\mathcal{E}^{(v)}|_2$ denotes the Hilbert-Schmidt norm of $\mathcal{E}^{(v)}$. This relation of stress and strain models a generalized Newtonian or a power-law-type fluid.

In this direction, only few results are known so far. The theory of weak solutions for the Navier-Stokes equations on power-law-type fluids on fixed domains was initiated by Ladyzhenskaya in [Lad69] and developed by Nečas, Málek, Ružička, Frehse and their co-authors, cf. for example the survey in [MR05]. The existence of strong solutions was studied in [DR05] for power-law-type fluids of exponent $d > \frac{7}{5}$. Results on the whole space case are given in [Wie05]. The fluid model introduced above which includes power-law-type fluids of exponent up to $d \geq 1$ was introduced in [BP07]. The authors show that a unique local strong L^p-solution exists on domains with compact boundary. In [Din07], this result was used to show the existence of strong solutions for the generalized Navier-Stokes problem in the exterior of several obstacles moving with a prescribed time-dependent velocity.

In [GV01], [GVP+02] and [Gal02, Part II], the steady motion and possible orientations of particles in second-order liquids, cf. [Jos90, Chapter 17], was studied. Interesting phenomena can be observed as in these viscoelastic fluids, inertia and normal stress "compete" to determine the particle's orientation.

The existence of weak solutions to the fully coupled system was proved recently in [FHN08], provided that the fluid is of shear-thickening power-law-type. We refer to Section 5.1 for more details.

As a second main result of this thesis, we prove the existence of a unique local strong L^p-solution for the coupled generalized Newtonian model, cf.

Theorem 5.1. It does not strictly generalize the result in the Newtonian case but requires more regularity of the data.

This thesis is organized as follows.

In Chapter 1, we present basic notation and concepts for the function spaces which occur. Maximal regularity results on the Stokes and generalized Stokes problem are stated and we quote preliminary results on embedding and trace estimates and on the Bogovskiĭ operator.

In Chapters 2 through 4, the proof of the main result on Newtonian liquids is given. In Chapter 2, we perform a change of coordinates on the model equations in order to rewrite them on a fixed domain. For the fluid part of the problem, this is similar to the transform in [IW77]. The rigid-body equations also change and become non-linear. It is important to note that in contrast to the situation of prescribed movement of the obstacle, the transform is an unknown part of the solution. Therefore, in order to do the fixed point argument for the non-linear problem, estimates on the transform in terms of possible body velocities are shown in Section 2.3.

Chapter 3 is devoted to the proof of maximal regularity of the linearized transformed problem on time intervals $(0, T)$, $T > 0$. In a way, we "outsource" the existence theory for the fluid part of the problem to known results on the Stokes problem with inhomogeneous Dirichlet boundary conditions, cf. Section 1.5. The remaining problem in the unknown body velocities is written as a linear fixed point equation in the space $W^{1,p}((0, T); \mathbb{R}^6)$, cf. Section 3.1. Its solvability depends on local pressure estimates for the Stokes problem which are adopted from [NS03], cf. Section 1.6, and on an appropriate representation of the forces which accelerate the rigid body, cf. [Gal02].

In Chapter 4, the transformed non-linear problem is solved by a contraction mapping argument. It relies on embeddings of the underlying spaces of maximal regularity for the linearized problem, cf. Section 1.4. They are needed in order to deal with the non-linear transformed convection and additional gradient terms. Furthermore, the results from Section 2.3 are used to give estimates on the highest order non-linearities which arise from the transform. The first main result of this thesis, the existence of a unique local strong L^p-solution for the free movement of a rigid body in a Newtonian fluid filling the exterior domain is stated precisely and proved in Section 4.3.

Chapter 5 is devoted to the L^p-theory in the case when the fluid is generalized Newtonian, i.e. its viscosity is non-constant and it satisfies (4). It is an advantage of our method introduced in Chapters 2-4 that it can be extended to this case, but also new arguments are necessary. The dependence of the viscosity on the shear rate implies that the operator A given by

$$A(v)_i := (\text{div}\,(\mu(|\mathcal{E}^{(v)}|_2^2)\mathcal{E}^{(v)}))_i$$

$$= \mu(|\mathcal{E}^{(v)}|_2^2)\Delta v_i + 2\mu'(|\mathcal{E}^{(v)}|_2^2)\sum_{j,k,l=1}^{3}\varepsilon_{ij}^{(v)}\varepsilon_{kl}^{(v)}\partial_j\partial_l v_k,$$

which replaces the Laplacian in the fluid equations, is quasi-linear. In [BP07], Bothe and Prüss showed that the fluid equations governed by A still yield maximal L^p-regularity. One main idea of the proof is to freeze A at a reference solution v_* and to consider the linear operator A_* given by

$$(A_*v)_i = \mu(|\mathcal{E}^{(v_*)}|_2^2)\Delta v_i + 2\mu'(|\mathcal{E}^{(v_*)}|_2^2)\sum_{j,k,l=1}^{3}\varepsilon_{ij}^{(v_*)}\varepsilon_{kl}^{(v_*)}\partial_j\partial_l v_k$$

instead. Under the above assumptions on μ and under the restriction that $p > 5$, the coefficients in the second-order differential operator A_* are sufficiently smooth to show maximal regularity for the corresponding fluid problem in view of [DHP07]. The condition $p > 5$ is due to the fact that coefficients are estimated by the norm of the velocity gradient ∇v and thus $v \in C^1(\overline{\mathcal{D}})$ is required. The result on $A(v)$ is then shown via a fixed point argument.

In view of the estimates and results from Chapters 2-4, two main tasks remain in order to prove the existence of a unique local strong L^p-solution for the fluid-rigid body interaction problem in the generalized Newtonian setting. The first is to show how the result by Bothe and Prüss can be used to solve the linearized transformed problem as in the Newtonian situation. In particular, additional local pressure estimates for this problem are needed, cf. Section 5.4. The second task is to verify that the difference of the transformed term $\mathcal{A}(u)$ of $A(v)$ and A_*u can be treated via a fixed point argument. This is done in Section 5.5.

Chapter 6 transfers the two main results on Newtonian and generalized Newtonian fluids to the cases of a bounded fluid domain and to the two-dimensional setting. It is shown that the structure of the arguments in the previous chapters can be preserved and that minor changes suffice to prove these extensions.

Acknowledgments

At this point, I would like to express my gratitude to several people who were involved in the writing of this thesis.

First of all, my thanks go to my advisor, Professor Matthias Hieber. He proposed the project of this thesis to me and I am grateful for the support he gave regarding the mathematical work and the problems and opportunities that go along with it.

I would also like to thank Professor Dieter Bothe and Professor Yoshihiro Shibata for acting as co-referees to this thesis. I am moreover grateful to Professor Shibata as he contributed to this project in a substantial way when he discussed and explained his ideas during his visit to Darmstadt in March 2009.

I would like to thank my colleagues in the research group of Professor Hieber, Bálint Farkas, Matthias Geißert, Robert Haller-Dintelmann, Tobias Hansel, Horst Heck, Matthias Heß, Okihiro Sawada, Verena Schmid and Kyriakos Stavrakidis. They "leave the doors of their offices open" and it is always possible to discuss and exchange on mathematics and other things.

I am grateful to Matthias Geißert for cooperating and discussing with me on this project. Robert Haller-Dintelmann helped to improve this thesis a lot in several ways. I would like to thank him especially for reading and correcting a preliminary version very thoroughly.

In addition, I would like to thank the Center of Smart Interfaces and the International Research Training Group 1529: Mathematical Fluid Dynamics for their financial support.

Zusammenfassung

Das Thema der vorliegenden Dissertation sind Systeme von Gleichungen, die die freie Bewegung eines Festkörpers in einer viskosen, inkompressiblen Flüssigkeit beschreiben. Sie sind auf dem bewegten Gebiet, das die Flüssigkeit füllt, durch die Navier-Stokes-Gleichungen gegeben. Die Geschwindigkeit des Schwerpunkts des Festkörpers und seine Winkelgeschwindigkeit sind über äußere Kräfte und Kräfte, die die Strömung der Flüssigkeit erzeugt, bestimmt. Umgekehrt wirkt die Geschwindigkeit des Festkörpers durch eine "no-slip" Bedingung an der Grenzfläche auf die Flüssigkeit zurück.

Das Hauptresultat dieser Arbeit ist die Existenz einer eindeutigen zeitlokalen starken L^p-Lösung für dieses gekoppelte Problem. Zum Beweis werden die Gleichungen in einem ersten Schritt durch eine Variablentransformation auf eine Formulierung in einem unbewegten Gebiet zurückgeführt. Mit Hilfe maximaler-Regularitätsabschätzungen für das transformierte und linearisierte System und eines Fixpunktarguments wird gezeigt, dass es genau eine Lösung des nichtlinearen transformierten Problems gibt. Die Hauptschwierigkeit liegt hier in der Behandlung der starken Kopplung der Bewegung von Festkörper und Flüssigkeit bei der Lösung des linearen Problems. Durch Rücktransformation folgt das Resultat.

In Kapitel 5 wird dieses Ergebnis von Newton'schen auf eine Klasse von nicht-Newton'schen Flüssigkeiten verallgemeinert, unter schärferen Voraussetzungen an die Regularität der Daten.

Die Beweismethode lässt sich darüber hinaus von dem Fall, dass die Flüssigkeit das gesamte Außenraumgebiet um den Körper füllt, auch auf beschränkte Gebiete übertragen und liefert ähnliche Ergebnisse für das Problem in zwei Raumdimensionen.

Chapter 1

Preliminaries

In this chapter, we introduce basic notation and preliminary concepts which will be used in the remainder of the thesis. Except for Section 1.1, the focus is on the Stokes and generalized Stokes problem and, more specifically, on the existence and properties of strong solutions in the L^p-setting for the equations on bounded and exterior domains.

1.1 Notation and Function Spaces

Most of the notation we introduce is standard. In this section we also briefly state known interpolation and trace results on the function spaces which occur later on. We always use generic constants C which may change from line to line. Their dependence on parameters is expressed only if necessary.

Let m, n, k be natural numbers and let Ω be a domain in $\mathbb{R}^n$. For every $x_0 \in \mathbb{R}^n$, the open ball with radius $R > 0$ centered at x_0 is denoted by $B_R(x_0) := \{x \in \mathbb{R}^n : |x - x_0| < R\}$ and we use the short notation $B_R := B_R(0)$.

For a function $f : \Omega \to \mathbb{R}^n$, $D^k f$ denotes its k-th derivative. Furthermore, given a multi-index $\alpha = (\alpha_1, \alpha_2, \ldots, \alpha_n) \in \mathbb{N}_0^n$ of size $|\alpha| = \Sigma_{i=1}^n \alpha_i$ we use the abbreviation $\partial^\alpha f := \partial_1^{\alpha_1} \partial_2^{\alpha_2} \ldots \partial_n^{\alpha_n} f$, where $\partial_i f$ denotes the partial derivative of f with respect to the i-th argument. It is used in the classical as well as in the distributional sense.

The spaces of all k-times continuously differentiable functions are denoted by $C^k(\Omega; \mathbb{R}^n)$ or simply $C^k(\Omega)$ if the dimension is clear from the context. This simplification applies to all spaces of $\mathbb{R}^n$-valued functions which will

1

appear, in order to keep the notation shorter. The space $C_c^k(\Omega)$ denotes the subspace of functions in $C^k(\Omega)$ which are compactly supported. Furthermore, for $\lambda \in (0, 1]$ let

$$C^{k,\lambda}(\Omega) := \{f \in C^k(\Omega) : \sup_{x,y \in \Omega, x \neq y} \frac{|D^k f(x) - D^k f(y)|}{|x - y|^\lambda} < \infty\}$$

the spaces of k-times differentiable functions with Hölder-continuous derivatives. All these notations extend to $k = 0$ and $k = \infty$.

We say that Ω is a domain of class $C^{k,\lambda}$, if for every $x_0 \in \partial\Omega$ there exists a neighborhood U_{x_0} of x_0 and a bijective map φ from U_{x_0} onto the cube $Q := (-1, 1)^n$ such that $\varphi(U_{x_0} \cap \Omega) \subset \mathbb{R}_+^n$, $\varphi(U_{x_0} \cap \partial\Omega) \subset \partial\mathbb{R}_+^n$, $\varphi \in C^{k,\lambda}(U_{x_0}; Q)$ and $\varphi^{-1} \in C^{k,\lambda}(Q; U_{x_0})$, where $\mathbb{R}_+^n := \{x \in \mathbb{R}^n : x_n > 0\}$ denotes the upper half plane in $\mathbb{R}^n$.

The domains which appear in our model are of class $C^{2,1}$. This is sufficient for the embedding and interpolation properties we need for the function spaces defined below on Ω.

For $1 \leq q \leq \infty$, $L^q(\Omega)$ denotes the Lebesgue spaces of $\mathbb{R}^n$- or $\mathbb{C}^n$-valued functions and $L^q(\Omega; X)$ denotes the Lebesgue spaces of functions which take values in a Banach space X. We use the short notation $\|f\|_q := \|f\|_{L^q(\Omega)}$ if no misinterpretation is possible.

For $T > 0$, the open interval $(0, T)$ is denoted by J_T. The spaces $L^p(J_T; L^q(\Omega))$ will occur many times and therefore we abbreviate their norms by $\|\cdot\|_{p,q} := \|\cdot\|_{L^p(J_T; L^q(\Omega))}$. Furthermore, we use the notation $L_{\text{loc}}^q(\Omega)$ for the spaces of functions f which satisfy $f|_K \in L^q(K)$ for every compact $K \subset \Omega$ and we let q' be the Hölder conjugated exponent of q satisfying $\frac{1}{q} + \frac{1}{q'} = 1$.

The Sobolev spaces of order m are denoted by $W^{m,q}(\Omega)$ with norm

$$\|f\|_{W^{m,q}(\Omega)} = \Big(\sum_{|\alpha| \leq m} \|\partial^\alpha f\|_q^q\Big)^{1/q}.$$

The closure of $C_c^\infty(\Omega)$ with respect to this norm is denoted by $W_0^{m,q}(\Omega)$. The homogeneous Sobolev spaces of equivalence classes of functions in $D^{m,q} := \{f \in L_{\text{loc}}^1(\Omega) : \partial^\alpha f \in L^q(\Omega), |\alpha| = m\}$ with respect to the polynomials of degree $m - 1$ are denoted by $\widehat{W}^{m,q}(\Omega)$. They are Banach spaces with the norms

$$\|f\|_{\widehat{W}^{m,q}(\Omega)} = \Big(\sum_{|\alpha|=m} \int_\Omega |\partial^\alpha f(x)|^q \, dx\Big)^{1/q}.$$

We introduce two scales of intermediate spaces of $W^{m,q}(\Omega)$. Bessel potential spaces appear as complex interpolation spaces in the context of fractional powers of generating operators, cf. Section 1.4. Besov and Sobolev-Slobodeckiĭ spaces are defined by real interpolation and occur as time and spatial trace spaces.

In the following, let X, Y be two Banach spaces. The space of continuous linear mappings from X to Y is denoted by $\mathcal{L}(X, Y)$ and we put $\mathcal{L}(X) := \mathcal{L}(X, X)$.

If X, Y form an interpolation couple, the complex interpolation space of X and Y is denoted by $[X, Y]_\theta$ for $\theta \in (0, 1)$. For basic properties we refer to [Tri95, Section 1.9].

On a domain Ω of class C^m and for $1 < q < \infty$ and a real number $0 < s \leq m$, the Bessel potential spaces of order s are defined by complex interpolation,

$$W^{s,q}(\Omega) := [L^q(\Omega), W^{m,q}(\Omega)]_{\frac{s}{m}}$$

as in [AF03, 7.57]. They also have intrinsic characterizations, where fractional derivatives on $L^q(\mathbb{R}^n)$ can be understood via Fourier transform, see for example [AF03, 7.63 and 7.64]. Note that this definition is compatible with the Sobolev spaces of integer order on the above type of domains.

For $0 < \theta < 1$ and $1 \leq p \leq \infty$ we denote the real interpolation spaces of X, Y by $(X, Y)_{\theta,p}$, cf. [Tri95, 1.3.2].

For every $0 < s < m, 1 \leq q < \infty, 1 \leq p \leq \infty$, we define Besov spaces on general domains by real interpolation of Sobolev spaces,

$$B^s_{q,p}(\Omega) := (L^q(\Omega), W^{m,q}(\Omega))_{s/m,p},$$

as in [AF03, 7.32]. The Sobolev-Slobodeckiĭ spaces $W^s_q(\Omega)$ for $0 < s < m$, $1 < q < \infty$ are given by

$$W^s_q(\Omega) := \begin{cases} W^{s,q}(\Omega) & \text{if } s \in \mathbb{N}, \\ B^s_{q,q}(\Omega) & \text{if } s \notin \mathbb{N}. \end{cases} \tag{1.1}$$

For $s \notin \mathbb{N}$, they coincide with the Bessel potential spaces if $q = 2$, cf. [Tri95, 4.6.1 (2)].

For a sufficiently regular function f on a Lipschitz domain Ω, we denote the trace f on $\partial\Omega$ by γf. A classical trace theorem on Sobolev spaces states that for every $f \in W^{1,q}(\Omega)$,

$$\|\gamma f\|_{L^r(\partial\Omega)} \leq C \|f\|_{L^q(\Omega)}^{(1-\lambda)} \|f\|_{W^{1,q}(\Omega)}^{\lambda} \tag{1.2}$$

if Ω is a domain of class $C^{1,1}$ and if $r = q(n-1)/(n-q)$ for $q < n$ or $r \in [1, \infty)$ for $q > n$ and $\lambda = n(r-q)/q(r-1)$, cf. [Gal94, Thm. II.3.1] or [AF03, Thm. 5.36]. This embedding carries over to the spaces $\widehat{W}^{1,q}(\Omega)$ on exterior domains Ω since the boundary is compact and we have $u \in L^q_{\mathrm{loc}}(\Omega)$ for all $u \in \widehat{W}^{1,q}(\Omega)$.

We use the trace embedding theorem for Besov spaces, cf. e.g. [AF03, Thm. 7.43], in order to get the following specific estimate.

Proposition 1.1. *For $1 < q < \infty$ and on a domain Ω of class C^1 the embedding*

$$W^{1/q+\varepsilon,q}(\Omega) \hookrightarrow L^q(\partial\Omega)$$

is continuous for all $0 < \varepsilon \leq 1 - \frac{1}{q}$.

Proof. The constant $\varepsilon > 0$ is used up for the limiting case $L^q(\partial\Omega)$ and in order to get embeddings from Bessel potential spaces. By [Tri95, 4.6.1], $W^{1/q+\varepsilon,q}(\Omega) \hookrightarrow B^{1/q+\varepsilon}_{q,2}(\Omega)$ for all $1 < q \leq 2$, $W^{1/q+\varepsilon,q}(\Omega) \hookrightarrow B^{1/q+\varepsilon}_{q,q}(\Omega)$ if $2 \leq q < \infty$ and $B^{1/q+\varepsilon}_{q,p}(\Omega) \hookrightarrow B^{1/q}_{q,1}(\Omega)$ for all $1 < p < \infty$. In [AF03, 7.44 and 7.45] it is shown hat $B^{1/q}_{q,1}(\mathbb{R}^n) \hookrightarrow L^q(\mathbb{R}^{n-1})$ for all $n \in \mathbb{N}$ and that this property extends to domains Ω of class C^1. $\qquad\square$

For $T > 0$ we use the special notation

$$_0W^{1,p}(J_T) := \{f \in W^{1,p}(J_T) : f(0) = 0\} \tag{1.3}$$

for the subspace of functions which can be extended to zero by a zero trace.

1.2   The Helmholtz Decomposition

In this section, we give some properties of the Helmholtz decomposition of a vector-valued function in $L^q(\Omega)$, $1 < q < \infty$, into a solenoidal and a potential part. This notion is of great help in the study of the Navier-Stokes equations and as part of the definition of the Stokes operator. For a detailed discussion of the Helmholtz decomposition and proofs of the subsequent properties, we refer to [Gal94]. The vector space of solenoidal fields in $L^q(\Omega)$ is denoted by

$$L^q_\sigma(\Omega) := \overline{C^\infty_{c,\sigma}(\Omega)}^{\|\cdot\|_q},$$

where $C_{c,\sigma}^{\infty}(\Omega) := \{u \in C_c^{\infty}(\Omega) : \operatorname{div} u = 0\}$. It is a closed subspace of $L^q(\Omega)$ and hence a Hilbert space if $q = 2$. In this case we define $G_2(\Omega) := L_\sigma^2(\Omega)^\perp$ as the orthogonal complement of $L_\sigma^2(\Omega)$ and obtain the *Helmholtz decomposition*

$$L^2(\Omega) = L_\sigma^2(\Omega) \oplus G_2(\Omega)$$

of $L^2(\Omega)$. It implies the existence of a linear, bounded and orthogonal projection $P_{\Omega,2} : L^2(\Omega) \to L_\sigma^2(\Omega)$, which is called *Helmholtz projection*.

This method does not automatically apply in arbitrary $L^q(\Omega)$-spaces because the subspace $L_\sigma^q(\Omega)$ may not be complemented. Instead, the existence of the Helmholtz decomposition in $L^q(\Omega)$, $q \neq 2$, depends on the domain Ω. Its existence follows if it can be shown that the weak Neumann problem

$$(\nabla p, \nabla \phi) = (u, \nabla \phi), \qquad \phi \in \widehat{W}^{1,q'}(\Omega), \tag{1.4}$$

has a unique solution $p \in \widehat{W}^{1,q}(\Omega)$ for every $u \in L^q(\Omega)$ and vice versa. If the Helmholtz decomposition

$$L^q(\Omega) = L_\sigma^q(\Omega) \oplus G_q(\Omega) \tag{1.5}$$

holds, the complement $G_q(\Omega)$ of $L_\sigma^q(\Omega)$ in $L^q(\Omega)$ is given by

$$G_q(\Omega) := \{u = \nabla p : p \in \widehat{W}^{1,q}(\Omega)\}.$$

In the following proposition, we collect properties of the Helmholtz decomposition which are needed later on.

Proposition 1.2. *Let $\Omega \subset \mathbb{R}^n$ be a bounded or exterior domain of class C^2. Then (1.5) holds for $L^q(\Omega)$, $1 < q < \infty$. The associated Helmholtz projections $P_{\Omega,q} : L^q(\Omega) \to L_\sigma^q(\Omega)$ are linear and bounded and satisfy $P'_{\Omega,q} = P_{\Omega,q'}$. Moreover, $L_\sigma^q(\Omega)' = L_\sigma^{q'}(\Omega)$ and $G_q(\Omega)' = G_{q'}(\Omega)$.*

Additionally, there are two more characterizations of $L_\sigma^q(\Omega)$,

$$L_\sigma^q(\Omega) = \{u \in L^q(\Omega) : (u, \nabla p) = 0, \text{ for all } p \in \widehat{W}^{1,q'}(\Omega)\}$$

for any domain $\Omega \subset \mathbb{R}^n$, $n \geq 2$, and

$$L_\sigma^q(\Omega) = \{u \in L^q(\Omega) : \operatorname{div} u = 0, u \cdot \nu|_{\partial\Omega} = 0\} \tag{1.6}$$

for any locally Lipschitzian domain $\Omega \subset \mathbb{R}^n$, $n \geq 2$, where ν denotes the outer normal on $\partial\Omega$ and where we understand $\operatorname{div} u$ and the trace $u \cdot \nu|_{\partial\Gamma}$ in the sense of the generalized Gauss Theorem on Ω, see [Gal94, Lemma III.2.1 and p. 119]. We will make use of this characterization frequently.

1.3   Maximal Regularity Results

A main ingredient of our proof of the existence of strong solutions for the fluid-rigid body interaction problem is maximal regularity of the underlying fluid operators. For the Stokes operator, this is a classical result by Solonnikov we quote in Subsection 1.3.1. The corresponding result for a class of generalized Newtonian fluids was obtained by Bothe and Prüss, see Subsection 1.3.2.

For a precise definition of maximal L^p-regularity, consider the abstract Cauchy problem

$$\begin{cases} u'(t) - Au(t) & = & f(t), \quad t \in J_T, \\ u(0) & = & 0, \end{cases} \tag{1.7}$$

where $(A, D(A))$ is a closed, densely defined linear operator in some Banach space X with domain $D(A)$ and $f \in L^p(J_T; X)$, $p \in (1, \infty)$ and $T \in \mathbb{R}_+$.

Definition 1.3. Under the above assumptions, the operator A is said to admit *maximal L^p-regularity* on J_T in X, if for every $f \in L^p(J_T; X)$ there exists a unique *strong solution* u of (1.7).

The function u is a strong solution of (1.7), if and only if $u \in W^{1,p}(J_T; X) \cap L^p(J_T; D(A))$, u has vanishing trace at time zero and u satisfies (1.7) for almost all $t \in J_T$.

In the following, we use the short form

$$X_p^T := W^{1,p}(J_T; X) \cap L^p(J_T; D(A))$$

for the space of strong solutions of the abstract Cauchy problem which is also called the space of maximal regularity of the operator A.

Let now $Z_p := \{u(0) : u \in X_p^T\}$ be the time trace space of X_p^T. In [Ama95, Section III.4.10] it is shown that if A admits maximal L^p-regularity on an interval J_T, $T > 0$, Z_p can be characterized by

$$Z_p = (X, D(A))_{1-1/p, p}$$

and that X_p^T admits the continuous embedding

$$X_p^T \hookrightarrow C([0, T]; Z_p).$$

For the initial value problem

$$\begin{cases} u'(t) - Au(t) & = & f(t), \quad t \in J_T, \\ u(0) & = & u_0, \end{cases} \tag{1.8}$$

we get the following result.

Proposition 1.4. *Suppose that A admits maximal L^p-regularity on J_{T_0}, $T_0 > 0$, then for every $u_0 \in Z_p$ and $T \in J_{T_0}$ there exists a unique solution $u \in X_p^T$ of (1.8). In addition, there exists a constant C independent of T, f, u_0, such that*

$$\|u\|_{X_p^T} \leq C(\|f\|_{L^p(J_T;X)} + \|u_0\|_{Z_p}).$$

In the next subsection, we define the Stokes operator and show that Proposition 1.4 can be applied to the Stokes problem.

Throughout this thesis, we also use the term "maximal regularity" with a different meaning applying not to closed operators but systems of equations. For example, the full linear system we study in Chapter 3 will not be rewritten as an abstract Cauchy problem, but we can fit the notion of "strong solution" to this situation.

1.3.1 The Stokes Operator

For $1 < q < \infty$, the Stokes operator A_q with Dirichlet boundary conditions in $L_\sigma^q(\Omega)$ is defined by

$$\begin{cases} A_q u & := & P_{\Omega,q}\Delta u, \\ D(A_q) & := & W^{2,q}(\Omega) \cap W_0^{1,q}(\Omega) \cap L_\sigma^q(\Omega). \end{cases}$$

We denote the corresponding space of maximal regularity of A_q by

$$X_{p,q,\sigma}^T := W^{1,p}(J_T; L_\sigma^q) \cap L^p(J_T; D(A_q))$$

and the space for the associated pressure by

$$Y_{p,q}^T := L^p(J_T; \widehat{W}^{1,q}(\Omega)).$$

The following proposition is a classical result due to Solonnikov [Sol77].

Proposition 1.5. *Let $\Omega \subset \mathbb{R}^n$, $n \geq 2$, be a domain of class C^2 and $1 < p < \infty$, $0 < T < T_0$, $f \in L^p(J_T; L_\sigma^q(\Omega))$ and $u_0 \in Z_{p,q} := (L_\sigma^q(\Omega), D(A_q))_{1-\frac{1}{p},p}$. Then there exists a unique solution $u \in X_{p,q,\sigma}^T$ to the inhomogeneous Stokes problem*

$$\begin{cases} u'(t) - A_q u(t) & = & f(t), \quad t \in J_T, \\ u(0) & = & u_0, \end{cases}$$

and there exists a constant $C > 0$ independent of T, u_0 and f, such that

$$\|u\|_{X_{p,q,\sigma}^T} \leq C(\|f\|_{p,q} + \|u_0\|_{Z_{p,q}}).$$

Moreover, setting $\nabla p := (\mathrm{Id} - P_{\Omega,q})\Delta u$ it follows that $(u,p) \in X^T_{p,q,\sigma} \times Y^T_{p,q}$ solves

$$\begin{cases} u'(t) - \Delta u(t) + \nabla p(t) &= f(t), \quad t \in J_T, \\ u(0) &= u_0 \end{cases} \tag{1.9}$$

and satisfies the estimate

$$\|u\|_{X^T_{p,q,\sigma}} + \|p\|_{Y^T_{p,q}} \leq C(\|f\|_{p,q} + \|u_0\|_{Z_{p,q}}). \tag{1.10}$$

A more explicit characterization of the trace space $(L^q_\sigma(\Omega), D(A_q))_{1-\frac{1}{p},p}$ is given by Amann in [Ama00]. Let ν be the outer normal of Ω and define

$$B^s_{q,p,\sigma,0}(\Omega) := \begin{cases} \{f \in B^s_{q,p}(\Omega) : \mathrm{div}\, f = 0, f|_{\partial\Omega} = 0\}, & \text{if } 1/q < s \leq 2, \\ \{f \in B^s_{q,p}(\Omega) : \mathrm{div}\, f = 0, f \cdot \nu|_{\partial\Omega} = 0\}, & \text{if } 0 \leq s < 1/q, \end{cases}$$

where the conditions on the divergence and the boundary have to be understood in the distributional sense, as in the characterization (1.6) of $L^q_\sigma(\Omega)$. It is shown in [Ama00, Thm. 3.4] that

$$Z_{p,q} = B^{2-2/p}_{q,p,\sigma,0}.$$

We summarize the proof of this characterization as follows. Let $H_q(\Omega) := W^{2,q}(\Omega) \cap W^{1,q}_0(\Omega)$ be the space of functions $f \in W^{2,q}(\Omega)$ such that $f|_{\partial\Omega} = 0$ and the domain of the Dirichlet-Laplace operator Δ_q in $L^q(\Omega)$. The main step is to show that $L^q_\sigma(\Omega)$ can be separated from the interpolation, i.e.

$$Z_{p,q} = (L^q(\Omega), H_q(\Omega))_{1-1/p,p} \cap L^q_\sigma(\Omega). \tag{1.11}$$

The idea of the proof is to show that the operator $Q_1 \in L(H_q, D(A_q))$ given by $Q_1 u = (\mu + A_q)^{-1} P_{\Omega,q}(\mu + \Delta_q)u$ for some $\mu \in \rho(A_q)$ can be extended to a bounded projection Q_0 from $L^q(\Omega)$ to $L^q_\sigma(\Omega)$, so that (1.11) follows from abstract interpolation theory, see [Tri95, 1.17.1, Thm. I], [Ama00, Lemma 3.2] and [FM70, Thm. 1.1]. By interpolation theory, the boundary condition in the remaining trace space $(L^q(\Omega), H_q(\Omega))_{1-1/p,p}$ for the Dirichlet-Laplacian is preserved if $\frac{1}{q} < 2 - \frac{2}{p}$, i.e. $(L^q(\Omega), H_q(\Omega))_{1-1/p,p} = \{f \in B^{2-2/p}_{q,p}(\Omega) : u|_{\partial\Omega} = 0\}$ and it gets lost for $2 - \frac{2}{p} < \frac{1}{q}$, so that $(L^q(\Omega), H_q(\Omega))_{1-1/p,p} = B^{2-2/p}_{q,p}(\Omega)$, cf. e.g. [Tri95, 4.3.3].

1.3.2   The Generalized Stokes Operator

This subsection is relevant to the generalization of our main result on the free-fall problem to a particular class of fluids, which may be non-Newtonian.

Instead of the Stokes operator, we look at a wider class of second order differential operators which satisfy a special type of ellipticity condition.

Let $A(x, D)$ be a differential operator

$$(A(x, D)v(x))_i := \sum_{j,k,l=1}^n a_{ij}^{kl}(x) D_k D_l v_j(x), \quad x \in \Omega,$$

where $D_j := -i\partial_j$ for $j \in \{1, \dots, n\}$ and v is a $\mathbb{C}^n$-valued function on Ω. The operator $A(x, D)$ is called *normally elliptic*, if the spectrum $\sigma(A(x, \xi))$ of its symbol $A(x, \xi) = \Sigma_{k,l=1}^n a_{ij}^{kl}(x)\xi_k\xi_l$ is contained in the open right half-plane $\mathbb{C}_+$ for every $x \in \overline{\Omega}$ and $\xi \in \mathbb{R}^n$ with $|\xi| = 1$. This condition on the coefficients of the second-order operator A is necessary to prove maximal regularity, cf. [DHP07]. The operator is called *strongly elliptic*, if additionally for every $x \in \overline{\Omega}$ and $\xi \in \mathbb{R}^n$ with $|\xi| = 1$, the numerical range of $A(x, \xi)$ is a subset of $\mathbb{C}_+$.

Here, we want to consider a parabolic problem given by A which includes an additional pressure term and incompressibility condition. In order for the Lopatinskiĭ-Shapiro condition to be satisfied for this problem, additional requirements on the coefficients of A are made. This leads to the following definition.

Definition 1.6. The operator $A(x, D)$ is called *strongly normally elliptic* if it is strongly elliptic and if for each $x \in \Omega$,

$$\operatorname{Re} \sum_{i,j,k,l=1}^n a_{ij}^{kl}(x)(\xi_l u_j - \nu_l v_j)\overline{(\xi_k u_i - \nu_k v_i)} > 0$$

for all $\xi, \nu \in \mathbb{R}^n$, $|\xi| = |\nu| = 1$, $(\xi, \nu) = 0$, $u, v \in \mathbb{C}^n$, $\operatorname{Im}(u, v) \neq 0$.

For a detailed discussion of this notion, we refer to [BP07, Section 3].

Let now $A(t, x, D)$ be a second-order differential operator with time-dependent $\mathbb{R}^n \times \mathbb{R}^n$-valued coefficients $A^{kl}(t, x)$ and assume that $A(t, x, D)$ is strongly normally elliptic for every $t \in J_T$, $T > 0$. To formulate the main result on $A(t, x, D)$, the additional requirements on the regularity of the coefficients are:

1. the A^{kl} are continuous on $J_T \times \overline{\Omega}$,

2. the limit $\lim_{|x|\to\infty} A^{kl}(t, x)$ exists uniformly in $t \in J_T$ if Ω is unbounded and

3. the operator $A(t, \infty, D)$ given by the limit coefficients $A^{kl}(t, \infty) = \lim_{|x| \to \infty} A^{kl}(t, x)$ is strongly elliptic.

Proposition 1.7. *Let $\Omega \subset \mathbb{R}^n$, $n \geq 2$, be a domain of class $C^{2,1}$ with compact boundary $\partial\Omega$ and outer normal ν, $T > 0$ and $A(t, x, D)$ as above. Suppose that $p > 3$, $f \in L^p(J_T; L^p(\Omega))$ and $u_0 \in W_p^{2-2/p}(\Omega)$, $\mathrm{div}\, u_0 = 0$, $u_0|_{\partial\Omega} = h(0)$, where*

$$h \in W_p^{1-1/2p}(J_T; L^p(\partial\Omega)) \cap L^p(J_T; W_p^{2-1/p}(\partial\Omega))$$

and $h \cdot \nu|_{\partial\Omega} = 0$. Then the problem

$$\begin{cases} u_t + Au + \nabla p & = & f & in\ J_T \times \Omega, \\ \mathrm{div}\, u & = & 0 & in\ J_T \times \Omega, \\ u & = & h & on\ J_T \times \partial\Omega, \\ u(0) & = & u_0 & in\ \Omega, \end{cases} \qquad (1.12)$$

has a unique strong solution

$$(u, p) \in X_{p,p}^T(\Omega) \times Y_{p,p}^T(\Omega),$$

which depends continuously on the data in the corresponding function spaces.

This result is proved in [BP07, Theorem 4.1] in more generality. It can be modified to allow for different assumptions including an inhomogeneous divergence and full Dirichlet, Neumann and slip boundary conditions and $1 < p < 3, p \neq \frac{3}{2}$.

In Chapter 5, we apply this result to a class of operators which correspond to the generalized Newtonian setting of our problem. We refer to the solutions of (1.12) by using the notation

$$u = \mathcal{U}_A(f, h, u_0) \quad \text{and} \quad p = \mathcal{P}_A(f, h, u_0) \qquad (1.13)$$

for the operators which yield the unique velocity field and pressure for suitable data.

1.4   Embedding Properties of $X_{p,q}^T$

In the following, we put $0 < T < T_0$. In view of inhomogeneous boundary conditions for the fluid velocity, we define

$$X_{p,q}^T := W^{1,p}(J_T; L^q(\Omega)) \cap L^p(J_T; W^{2,q}(\Omega)),$$

as the space of fluid velocity solutions which includes $X_{p,q,\sigma}^T$ as a subspace. The following proposition yields embeddings of $X_{p,q}^T$ which are helpful in the estimates on the linear as well as the non-linear parts of our problem. In the subspace

$$X_{p,q,0}^T := \{u \in X_{p,q}^T : u|_{t=0} = 0\},$$

the embedding constants may be chosen independently of T.

Proposition 1.8. *Let $\Omega \subset \mathbb{R}^n$ be a $C^{1,1}$-domain with compact boundary, assume $p, q \in (1, \infty)$, $\alpha \in (0, 1)$ and fix $T_0 > 0$. Then*

$$X_{p,q}^{T_0} \hookrightarrow W^{\alpha,p}(J_{T_0}; W^{2(1-\alpha),q}(\Omega)).$$

In particular, if $r, s \in (1, \infty]$, $\mu \in \{0, 1\}$ and

$$\frac{2 - \mu}{2} + \frac{n}{2r} - \frac{n}{2q} \geq \frac{1}{p} - \frac{1}{s},$$

then for all $T \in (0, T_0]$, the embedding

$$X_{p,q}^T \hookrightarrow L^s(J_T; W^{\mu,r}(\Omega))$$

is continuous. Moreover, there exists a constant $C_0 = C(T_0)$, independent of $T \in J_{T_0}$, such that the estimate

$$\|u\|_{L^s(J_T; W^{\mu,r}(\Omega))} \leq C_0 \|u\|_{X_{p,q}^T}$$

holds true for all $u \in X_{p,q,0}^T$.

For a proof by the mixed derivatives theorem (see for example [DHP07]), we refer to [DGH09, Lemma 4.2].

At this point we also note the elementary embedding constants

$$\|f\|_p \leq T^{1/p - 1/q} \|f\|_q \qquad \text{for all } f \in L^q(J_T), q > p$$

and

$$\|f\|_\infty \leq T^{1/p'} \|f\|_{W^{1,p}(J_T)} \qquad \text{for all } f \in {}_0 W^{1,p}(J_T), \frac{1}{p} + \frac{1}{p'} = 1,$$

which will be used many times.

1.5 The Bogovskiĭ Operator and Inhomogeneous Dirichlet Boundary Conditions

The Bogovskiĭ operator B_Ω corresponding to a bounded domain Ω acts as a right inverse to the divergence operator. It gives a solution $u = B_\Omega g$ of the problem

$$\begin{cases} \operatorname{div} u &= g \quad \text{in } \Omega, \\ u|_{\partial\Omega} &= 0 \quad \text{on } \partial\Omega \end{cases}$$

on suitable function spaces and domains. More precisely, we cite the following result from [Bog79], [Gal94] and [GHH06b].

Proposition 1.9. *Let $\Omega \subset \mathbb{R}^n$ be a bounded domain with Lipschitz boundary. Then there exists an operator*

$$B_\Omega : C_c^\infty(\Omega) \to C_c^\infty(\Omega)^n$$

such that

$$\operatorname{div} B_\Omega g = g, \qquad \text{provided} \int_\Omega g = 0.$$

Moreover, for $1 < p < \infty$ and $s \geq 0$, B_Ω can be continuously extended to a bounded operator from $W_0^{s,p}(\Omega)$ to $W_0^{s+1,p}(\Omega)^n$.

In our situation, the Bogovskiĭ operator is used to deal with inhomogeneous Dirichlet boundary conditions for the Stokes problem. The idea is to extend the boundary data by a solenoidal vector field and then subtract it from the unknown velocity.

We now consider the following special case: Let $h \in W^{1,p}(J_T; C^2(\partial\Omega))$ be a function on the boundary of an exterior domain Ω of class $C^{2,1}$ such that there exists an extension H of h onto Ω satisfying

$$H|_{\partial\Omega} = h, \quad H \in W^{1,p}(J_T; C^2(\Omega)), \quad \text{and} \quad \operatorname{div} H = 0. \tag{1.14}$$

In particular, h and H could be given by a rigid motion $\xi + \Omega \times y$ on $\mathbb{R}^3$, where $\xi, \Omega \in W^{1,p}(J_T; \mathbb{R}^3)$. We choose open balls $B_1, B_2 \subset \mathbb{R}^n$ such that $\Omega^c \subset B_1 \subset \overline{B_1} \subset B_2$ and define a cut-off function $\chi \in C^\infty(\mathbb{R}^n; [0,1])$ satisfying

$$\chi(y) := \begin{cases} 1 & \text{if } y \in \overline{B_1}, \\ 0 & \text{if } y \in \Omega \setminus B_2. \end{cases} \tag{1.15}$$

Then we define

$$b_h := \chi H - B_K((\nabla\chi)H), \tag{1.16}$$

where $K \subset \mathbb{R}^n$ is a bounded open set which contains the annulus $B_1 \backslash \overline{B_2}$, so that $b_h \in W^{1,p}(0, T; C^2_{c,\sigma}(\mathbb{R}^n))$ due to $\operatorname{div} b_h = \nabla \chi H + \chi \operatorname{div} H - \nabla \chi H = 0$.

If we set $u := u_b + b_h$, the Stokes problem

$$
\begin{cases}
u_t - \Delta u + \nabla p &= f \quad \text{in } J_T \times \Omega, \\
\operatorname{div} u &= 0 \quad \text{in } J_T \times \Omega, \\
u|_{\partial \Omega} &= h \quad \text{on } J_T \times \partial \Omega, \\
u(0) &= u_0 \quad \text{in } \Omega,
\end{cases}
\tag{1.17}
$$

is equivalent to

$$
\begin{cases}
\partial_t u_b - \Delta u_b + \nabla p &= f + \Delta b_h - \partial_t b_h \quad \text{in } J_T \times \Omega, \\
\operatorname{div} u_b &= 0 \qquad\qquad\qquad\;\; \text{in } J_T \times \Omega, \\
u_b|_{\partial \Omega} &= 0 \qquad\qquad\qquad\;\; \text{on } J_T \times \partial \Omega, \\
u_b(0) &= u_0 - b_h(0) \qquad\;\; \text{in } \Omega.
\end{cases}
\tag{1.18}
$$

By (1.10), we have a solution (u_b, p) of (1.18) if $u_0 - b_h(0) \in Z_{p,q}$. The functions u and p thus solve (1.17) and they satisfy

$$
\|u\|_{X^T_{p,q}} + \|p\|_{Y^T_{p,q}} \leq C(\|f\|_{p,q} + \|u_0 - b_h(0)\|_{Z_{p,q}} + \|b_h\|_{X^T_{p,q}})
\tag{1.19}
$$

by Proposition 1.5 and (1.10). As in the generalized Newtonian case, we define solution operators

$$
\mathcal{U}(f, h, u_0) \in X^T_{p,q}, \qquad \mathcal{P}(f, h, u_0) \in Y^T_{p,q}
\tag{1.20}
$$

for the inhomogeneous problem (1.17).

1.6 Local Pressure Estimates

If the exterior force in the inhomogeneous Stokes problem is solenoidal, the pressure term inherits regularity from the velocity on bounded domains, where the Poincaré inequality can be applied. In the following lemma, we transfer these estimates proved by Noll and Saal in [NS03, Lemma 13] for the Stokes resolvent problem to the parabolic situation, in which we can make use of Proposition 1.8 to convert between space and time regularity.

On bounded domains Ω we define

$$
L^q_0(\Omega) := \{ f \in L^q(\Omega) : \int_\Omega f = 0 \}
$$

for $1 < q < \infty$.

Lemma 1.10. *Let $\Omega \subset \mathbb{R}^n$, $n \geq 2$, be an exterior domain, $1 < p, q < \infty$, $0 < T < T_0$, and $f \in L^p(J_T; L^q_\sigma(\Omega))$. Then if the pressure part $p = \mathcal{P}(f, 0, 0) \in Y^T_{p,q}$ of the solution of (1.17) with zero initial and boundary conditions is chosen in a way that $p \in L^p(J_T; L^q_0(\Omega_R))$ for some $R > 0$, $\Omega_R := \Omega \cap B_R$, it satisfies the estimate*

$$\|p\|_{L^p(J_T; L^q(\Omega_R))} \leq CT^{\alpha/p} \|f\|_{p,q} \qquad \text{for all } 0 \leq \alpha < \frac{1}{2q}.$$

Proof. Choose Ω_R and p from $L^p(J_T; \widehat{W}^{1,q}(\Omega))$ such that $\partial\Omega \subset \partial\Omega_R$ and $p \in L^p(J_T; L^q_0(\Omega_R))$. Let $u = \mathcal{U}(f, 0, 0)$ be the velocity field p corresponds to, then

$$\nabla p(t, x) = ((\mathrm{Id} - P_{\Omega,q})\Delta u(x))(t) \quad \text{for a. a. } t \in J_T.$$

Using a similar argument as in the proof of [NS03, Lemma 13], for almost all $t \in J_T$, we obtain

$$\|p(t)\|_{L^q(\Omega_R)} \leq C \|u(t)\|_{W^{2-2\alpha,q}(\Omega)}$$

for all $0 \leq \alpha < \frac{1}{2q}$. By interpolation, Proposition 1.8 and by (1.10),

$$
\begin{aligned}
\|p\|_{L^p(J_T; L^q(\Omega_R))} &\leq C \|u\|^{1-\alpha}_{L^p(J_T; W^{2,q}(\Omega))} \|u\|^{\alpha}_{p,q} \leq CT^{\alpha/p} \|u\|^{1-\alpha}_{X^T_{p,q,\sigma}} \|u\|^{\alpha}_{\infty,q} \\
&\leq CT^{\alpha/p} \|u\|_{X^T_{p,q,\sigma}} \leq CT^{\alpha/p} \|f\|_{p,q}, \qquad\qquad (1.21)
\end{aligned}
$$

where the constant C may be chosen independently of $T < T_0$. $\qquad\square$

The estimates in Chapter 5 demand a similar result for the generalized Stokes problem, which will be proved in Subsection 5.4.1.

Chapter 2

Mathematical Model and Change of Coordinates

The aims of this chapter are to introduce the mathematical formulation of the fluid-rigid body interaction problem and the change of coordinates which enables us to rewrite the problem on a fixed domain. The third section of this chapter is devoted to several technical estimates regarding this change of coordinates which will be important for the fixed point arguments in Chapter 4 and in Section 5.5.

2.1 Mathematical Formulation of the Free-Fall Problem

In $\mathbb{R}^3$, the rigid body occupies a bounded domain $\mathcal{B}(t)$ with boundary $\Gamma(t)$ and outer normal $n(t)$. Additionally, it is required that its complement in $\mathbb{R}^3$ is a domain as well. The body moves with a rigid velocity

$$v_{\mathcal{B}}(t, x) = \omega(t) \times (x - x_c(t)) + \eta(t),$$

where $\eta(t) = \dot{x}_c(t) + \eta_0 \in \mathbb{R}^3$ is the translational velocity of its center of mass $x_c(t) \in \mathbb{R}^3$ and $\omega(t) \in \mathbb{R}^3$ is its angular velocity at time $t \geq 0$. For simplicity, we assume $x_c(0) = 0$. The body is completely immersed in a Newtonian fluid which fills the exterior domain $\mathcal{D}(t) = \mathbb{R}^3 \backslash \overline{\mathcal{B}(t)}$.

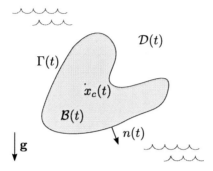

Up to any time $T > 0$, the system of equations governing the movement of both liquid and body is given by the Navier-Stokes equations on $J_T \times \mathcal{D}(\cdot) := \{(t, x) \in \mathbb{R}^4 : t \in J_T, x \in \mathcal{D}(t)\}$, and by the laws of conservation of inertia and angular momentum for the body,

$$
\left\{
\begin{aligned}
v_t - \operatorname{div} \mathbf{T}(v, q) + (v \cdot \nabla)v &= \mathbf{g} && \text{in } J_T \times \mathcal{D}(\cdot), \\
\operatorname{div} v &= 0 && \text{in } J_T \times \mathcal{D}(\cdot), \\
v &= v_\mathcal{B} && \text{on } J_T \times \Gamma(\cdot), \\
v(0) &= v_0 && \text{in } \mathcal{D}(0), \\
\mathrm{m}\eta' + \int_{\Gamma(t)} \mathbf{T}(v, q)n \, \mathrm{d}\sigma &= \mathbf{F} && \text{in } J_T, \\
(J\omega)' + \int_{\Gamma(t)} (x - x_c(t)) \times \mathbf{T}(v, q)n \, \mathrm{d}\sigma &= \mathbf{M} && \text{in } J_T, \\
\eta(0) &= \eta_0, \\
\omega(0) &= \omega_0.
\end{aligned}
\right.
\tag{2.1}
$$

Here, v, q denote the velocity and the scalar pressure of the fluid and $\mathbf{g}$ is a constant vector modeling gravitation. Without loss of generality, we set the fluid density to 1 but admit arbitrary density $\rho_\mathcal{B}$ and mass m of the body. The functions $\mathbf{F}$ and $\mathbf{M}$ indicate the total external force and the total external torque acting on the obstacle with respect to x_c. To model a free fall of the body, we set $\mathbf{F} = \mathrm{m}\mathbf{g}$, $\mathbf{M} = 0$, although additional external forces and torques are admissible.

The inertia tensor J of the body is given by the relation

$$
aJ(t)b = \int_{\mathcal{B}(t)} \rho_\mathcal{B}(x)(a \times (x - x_c(t))) \cdot (b \times (x - x_c(t))) \, \mathrm{d}x
$$

for all $a, b \in \mathbb{R}^3$. The movement of body and fluid is coupled in two ways. The equations on the body velocity include the Newtonian stress tensor

$$
\mathbf{T}(v, q) = 2\mu \mathcal{E}^{(v)} - q\mathrm{Id},
$$

where μ is the kinematic viscosity and

$$\mathcal{E}^{(v)} := \frac{1}{2}(\nabla v + (\nabla v)^T)$$

yields the strain tensor of the fluid. The fluid exerts a drag or lift force $-\int_{\Gamma(t)} \mathbf{T}(v,q)n\,d\sigma$ and a torque $-\int_{\Gamma(t)} (x - x_c) \times \mathbf{T}(v,q)n\,d\sigma$ on the body. On the other hand, the full velocity v_B of the body influences the fluid flow via Dirichlet no-slip conditions on the interface.

The way we introduced this model, the body is not necessarily dropped from rest and the fluid may have non-zero initial velocity. Apart from these minor changes, it coincides with Galdi's definition of the Free-Fall Problem in [Gal02].

One difficulty of the above system of equations is its setting on the unknown domain $\mathcal{D}(t)$. In the next subsection, we change coordinates and define new variables which give an equivalent formulation of the problem on the fixed domain $\mathcal{D} := \mathcal{D}(0)$.

2.2   Change of Coordinates

Let now $\mathcal{B} := \mathcal{B}(0)$ and $\Gamma := \Gamma(0)$. If we adjust our view to fixed points $y \in \mathcal{B}$, their positions at any time t can be described by an affine map

$$X_0(t,y) = Q(t)y + x_c(t),$$

where $Q(t)$ is an orthogonal matrix since the body can only perform rigid motions. A second characterization of this map on $\mathbb{R}^3$ through the body's velocities is given by the set of differential equations

$$\begin{cases} \partial_t X_0(t,y) &= \omega(t) \times (X_0(t,y) - x_c(t)) + \eta(t), \quad t \in \mathbb{R}_+, y \in \mathbb{R}^3, \\ X_0(0,y) &= y, \qquad\qquad\qquad\qquad\qquad\qquad\quad y \in \mathbb{R}^3, \end{cases} \quad (2.2)$$

which is parametrized by their initial conditions. Given $\eta, \omega \in W^{1,p}(J_T) \hookrightarrow C([0,T])$ on any time interval J_T, $T > 0$, its solutions have regularity $X_0 \in C^1(J_T; C^\infty(\mathbb{R}^3))$ by the Picard-Lindelöf Theorem and it is easy to show that we even get $X_0 \in W^{2,p}(J_T; C^\infty(\mathbb{R}^3))$. The corresponding inverse mappings $Y_0(t)$ of $X_0(t)$ are given by

$$Y_0(t,x) = Q^T(t)(x - x_c(t))$$

or the differential equations

$$\begin{cases} \partial_t Y_0(t,x) &= -\Omega(t) \times Y_0(t,x) - \xi(t), \quad t \in \mathbb{R}_+, x \in \mathbb{R}^3, \\ Y_0(0,x) &= x, \qquad\qquad\qquad\qquad\qquad x \in \mathbb{R}^3, \end{cases} \tag{2.3}$$

where

$$\Omega(t) := Q^T(t)\omega(t), \quad \xi(t) := Q^T(t)\eta(t) \tag{2.4}$$

are the new transformed body velocities. In the following, we often use the notation $m(t)$ or $M(t)$ for the skew-symmetric matrices satisfying

$$m(t)x = \omega(t) \times x \quad \text{and} \quad M(t)y = \Omega(t) \times y.$$

Note that $m(t) = Q(t)M(t)Q^T(t)$ for all $t \in J_T$ since

$$m(t)x = \omega(t) \times x = (Q(t)\Omega(t)) \times x = Q(t)(\Omega(t) \times Q^T(t)x) = Q(t)M(t)Q^T(t)x$$

for all $x \in \mathbb{R}^3$, where we used the relation

$$Q(a \times b) = Qa \times Qb \qquad \text{for all } a,b \in \mathbb{R}^3, \tag{2.5}$$

which holds if Q is orthogonal.

We modify the diffeomorphisms X_0, Y_0 such that they rotate and shift space only on a suitable open neighborhood of the rotating and translating body. We choose open balls $B_1, B_2 \subset \mathbb{R}^3$ such that $\overline{\mathcal{B}} \subset B_1 \subset \overline{B_1} \subset B_2$ and define a cut-off function $\chi \in C^\infty(\mathbb{R}^3; [0,1])$ satisfying

$$\chi(y) := \begin{cases} 1 & \text{if } y \in \overline{B_1}, \\ 0 & \text{if } y \in \mathcal{D} \setminus B_2, \end{cases} \tag{2.6}$$

and a time-dependent vector field $b : [0,T] \times \mathbb{R}^3 \to \mathbb{R}^3$ by

$$\begin{aligned} b(t,x) \quad := \quad & \chi(x - x_c(t))[m(t)(x - x_c(t)) + \eta(t)] \\ & - B_K(\nabla\chi(\cdot - x_c(t))m(t)(\cdot - x_c(t)))(x). \end{aligned} \tag{2.7}$$

Here, $B_K : C_c^\infty(K; \mathbb{R}) \to C_c^\infty(K; \mathbb{R}^3)$ indicates the Bogovskiĭ operator for functions defined on a bounded open set K which contains the annulus $\overline{B_2} \setminus B_1$ but does not intersect $\mathcal{B}$, cf. Proposition 1.9. The function $b(t)$ belongs to $C_c^\infty(\mathbb{R}^3)$ for all $t \in [0,T]$. Due to

$$\int_{\overline{B_2} \setminus B_1} \nabla\chi(y - x_c(t))m(t)(y - x_c(t)) \, dy = \int_{\overline{B_2} \setminus B_1} \chi(t, y - x_c(t)) \operatorname{tr} m(t) \, dy = 0,$$

the correction by the Bogovskiĭ term yields $\operatorname{div} b(t) = 0$ for all $t \in [0,T]$. Thus from $\eta, \omega \in W^{1,p}(J_T)$ we get $b \in W^{1,p}(J_T; C_{c,\sigma}^\infty(\mathbb{R}^3))$ and $b|_\Gamma = v_\mathcal{B}$. There is also a more explicit way of defining b, which yields the same solenoidal, boundary and regularity properties, given by

$$b(t, x) := -\frac{1}{2}\text{rot}\,(\chi[(|x|^2 - 2x \cdot x_c(t))\omega + \eta \times x]).$$

It seems simpler in this case, but the correction via the Bogovskiĭ operator can be applied in more general situations in which the interface velocity is not a rigid motion, cf. Section 1.5. We will make use of this fact in Chapter 3.

We now consider the ordinary differential equation

$$\begin{cases} \partial_t X(t, y) &= b(t, X(t, y)), \quad t \in J_T, y \in \mathbb{R}^3, \\ X(0, y) &= y, \qquad\qquad\quad y \in \mathbb{R}^3. \end{cases} \quad (2.8)$$

It yields a unique solution $X \in C^1(J_T; C^\infty(\mathbb{R}^n))$ by the Picard-Lindelöf Theorem. The solution has continuous mixed partial derivatives $\frac{\partial^{|\alpha|+1} X}{\partial t (\partial y_j)^\alpha}$, $\frac{\partial^{|\alpha|} X}{(\partial y_j)^\alpha}$, where $\alpha \in \mathbb{N}_0^3$ denotes a multi-index, cf. [Har64, Theorem V.4.1]. By uniqueness, the function $X(t, \cdot)$ is bijective and we denote its inverse by $Y(t, \cdot)$. Since $\text{div}\, b = 0$, Liouville's Theorem, cf. [Har64, Theorem IV.1.2], implies that the Jacobians J_X and J_Y are volume-preserving, i.e.

$$J_X(t, y)J_Y(t, X(t, y)) = \text{Id}_{\mathbb{R}^3} \text{ and } \det J_X(t, y) = \det J_Y(t, x) = 1. \quad (2.9)$$

Given X, the inverse transform Y satisfies the differential equations

$$\begin{cases} \partial_t Y(t, x) &= b^{(Y)}(t, Y(t, x)), \quad t \in J_T, x \in \mathbb{R}^3, \\ Y(0, x) &= x, \qquad\qquad\quad x \in \mathbb{R}^3, \end{cases} \quad (2.10)$$

where

$$b^{(Y)}(t, y) := -J_X^{-1}(t, y)b(t, X(t, y)). \quad (2.11)$$

Note that by this definition, $b^{(Y)}$ and Y obtain the same space and time regularity as b and X. Within the ball B_1, X, Y coincide with X_0, Y_0; whereas in the complement of $K \cup \overline{B_2}$, $\partial_t X(t, y) = \partial_t Y(t, x) = 0$ and so $X(t, y) = y$ and $Y(t, x) = x$.

In order to write the equations (2.1) on the cylindrical domain $J_T \times \mathcal{D}$, in addition to the transformed body velocities ξ, Ω we introduce

$$\begin{aligned} u(t, y) &:= J_Y(t, X(t, y))v(t, X(t, y)), \\ p(t, y) &:= q(t, X(t, y)), \\ \mathbf{G}(t, y) &:= J_Y(t, X(t, y))\mathbf{g}, \\ \mathcal{T}(u, p) &:= Q^T \mathbf{T}(Qu, p)Q. \end{aligned} \quad (2.12)$$

The new stress tensor $\mathcal{T}$ will only be used close to the body, on the ball B_1. We denote the outer normal of $\mathcal{B}$ by $N := n(0) = Q^T(t)n(t)$ for all $0 \leq t \leq T$. It follows that $\mathcal{T}$ satisfies

$$\int_{\Gamma(t)} \mathbf{T}(v,q)n(t)\,\mathrm{d}\sigma = Q\int_{\Gamma} \mathcal{T}(u,p)N\,\mathrm{d}\sigma$$

and

$$\int_{\Gamma(t)} (x - x_c(t)) \times \mathbf{T}(v,q)n(t)\,\mathrm{d}\sigma = Q\int_{\Gamma} y \times \mathcal{T}(u,p)N\,\mathrm{d}\sigma,$$

where we use that $\det Q = \det Q^T = 1$. From (2.2) we see that Q satisfies the equations

$$\dot{Q} = QM, \qquad Q(0) = \mathrm{Id}_{\mathbb{R}^3}. \tag{2.13}$$

With this ODE, we can determine Q for any given ξ, Ω and calculate the corresponding Eulerian velocities via $\eta = Q\xi$ and $\omega = Q\Omega$. Furthermore, we get $m\dot{Q}\xi = mQM\xi$, so that the new equations on the transformed translational velocity of the body reads

$$m\xi' = m\mathbf{G}(t,0) - \int_{\Gamma} \mathcal{T}(u,p)N\,\mathrm{d}\sigma - m\Omega \times \xi.$$

We again use the relation (2.5) to show that the transformed inertia tensor $I = Q^T J Q$ no longer depends on time since

$$
\begin{aligned}
aIb &= \int_{\mathcal{B}(t)} \rho_{\mathcal{B}}(x)(aQ^T \times (x - x_c)) \cdot (Qb \times (x - x_c))\,\mathrm{d}x \\
&= \int_{\mathcal{B}} \rho_{\mathcal{B}}(y)Q(Q^TaQ^T \times y) \cdot Q(b \times y)\,\mathrm{d}y = \int_{\mathcal{B}} \rho_{\mathcal{B}}(y)(a \times y) \cdot (b \times y)\,\mathrm{d}y.
\end{aligned}
$$

For the transformed equation on the angular velocity, we thus obtain

$$I\Omega' = \Omega \times (I\Omega) - \int_{\Gamma} y \times \mathcal{T}(u,p)N\,\mathrm{d}\sigma,$$

since $\dot{J}\omega = (\dot{Q}IQ^T + QI\dot{Q}^T)\omega = Q(MI + IM)\Omega = Q(\Omega \times (I\Omega))$ by (2.13).

It is more tedious to calculate the transformed fluid equations from X and Y, but they are already well-known from the work of Inoue and Wakimoto in [IW77], where the Navier-Stokes equations are considered on domains which move in a prescribed way. Our transform X satisfies the regularity assumptions and condition on preservation of volume Inoue and Wakimoto require if we start from prescribed $\eta, \omega \in W^{1,p}(J_T)$. The justification for this assumption is given by the regularity of the solution we find through the fixed point argument in Chapter 4.

Let

$$g^{ij} = \sum_{k=1}^{n} (\partial_i Y_k)(\partial_j Y_k) \qquad (2.14)$$

the metric contravariant tensor,

$$g_{ij} = \sum_{k=1}^{n} (\partial_i X_k)(\partial_j X_k) \qquad (2.15)$$

the metric covariant tensor and

$$\Gamma^i_{jk} = \sum_{l=1}^{n} g^{jk}(\partial_i g_{jl} + \partial_j g_{lk} - \partial_k g_{ij}) \qquad (2.16)$$

the Christoffel symbol corresponding to our change of coordinates. For the new transformed unknowns u, p, ξ and Ω we obtain the following system of fluid and body equations in $J_T \times \mathcal{D}$,

$$\begin{cases}
u_t + (\mathcal{M} - \mathcal{L})u = \mathbf{G} - \mathcal{N}(u) - \mathcal{G}p & \text{in } J_T \times \mathcal{D}, \\
\operatorname{div} u = 0 & \text{in } J_T \times \mathcal{D}, \\
u(t, y) = \xi(t) + \Omega(t) \times y, & t \in J_T, y \in \Gamma, \\
u(0) = v_0 & \text{in } \mathcal{D}, \\
m\xi' + m(\Omega \times \xi) = m\mathbf{G}(\cdot, 0) - \int_\Gamma T(u, p)N \, d\sigma & \text{in } J_T, \\
I\Omega' + \Omega \times (I\Omega) = -\int_\Gamma y \times T(u, p)N \, d\sigma & \text{in } J_T, \\
\xi(0) = \eta_0, \\
\Omega(0) = \omega_0.
\end{cases} \qquad (2.17)$$

Here, the operator $\mathcal{L}$ denotes the transformed Laplace operator and it is given by

$$\begin{aligned}
(\mathcal{L}u)_i \;\; := \;\; & \sum_{j,k=1}^{n} \partial_j(g^{jk}\partial_k u_i) + 2 \sum_{i,k,l=1}^{n} g^{kl}\Gamma^i_{jk}\partial_l u_j \\
& + \sum_{j,k,l=1}^{n} \left(\partial_k(g^{kl}\Gamma^i_{jl}) + \sum_{m=1}^{n} g^{kl}\Gamma^m_{jl}\Gamma^i_{km} \right) u_j.
\end{aligned} \qquad (2.18)$$

The convection term is transformed into

$$(\mathcal{N}u)_i \;\; := \;\; \sum_{j=1}^{n} u_j \partial_j u_i + \sum_{j,k=1}^{n} \Gamma^i_{jk} u_j u_k.$$

The transformed time derivative and the transformed gradient are given by

$$(\mathcal{M}u)_i \;\; := \;\; \sum_{j=1}^{n} \dot{Y}_j \partial_j u_i + \sum_{j,k=1}^{n} \left(\Gamma^i_{jk}\dot{Y}_k + (\partial_k Y_i)(\partial_j \dot{X}_k) \right) u_j$$

and

$$(\mathcal{G}p)_i \quad := \quad \sum_{j=1}^{n} g^{ij}\partial_j p,$$

respectively. It can be shown as in the proof of [IW77, Theorem 2.5] that the fluid part of (2.17) is equivalent to the fluid part of (2.1).

2.3   Estimates on the Coordinate Transform

In the following, we show some basic estimates on the transforms X and Y in terms of the transformed body velocities ξ and Ω. They will be needed for the fixed point argument in Chapter 4.

Let $T > 0$. We first retrace the construction of X and Y from given $\xi, \Omega \in W^{1,p}(J_T)$ in five small steps.

1. First, we determine $Q \in W^{2,p}(J_T; \mathbb{R}^{3\times3})$ by solving the system of ordinary differential equations (2.13), where the matrix-valued function M satisfies $M(t)x = \Omega(t) \times x$ for all $t \in J_T$ and $x \in \mathbb{R}^3$.

2. Secondly, we calculate the original body velocities $\eta = Q^T\xi$ and $\omega = Q^T\Omega$.

3. Define $b = \chi(\cdot - x_c)v_{\mathcal{B}} - B_K(\nabla\chi(\cdot - x_c)v_{\mathcal{B}})$ as in (2.7), where

$$v_{\mathcal{B}}(t,x) = \omega(t) \times (x - x_c(t)) + \eta(t)$$

and $x_c(t) = \int_0^t \eta(s)\,\mathrm{d}s$ are known from step (2).

4. Solve the equations (2.8) with right hand side b from step (3) to get X.

5. Define $b^{(Y)} := J_X^{-1}b(\cdot, X)$ as in (2.11) and solve (2.10) to get the inverse Y of X.

From this procedure, we get the following estimates.

Proposition 2.1. *Let* $T > 0$, $\xi_1, \xi_2, \Omega_1, \Omega_2 \in W^{1,p}(J_T)$. *Then* $X_i, Y_i \in C^1(J_T; C^\infty(\mathbb{R}^n))$ *such that*

$$\|\partial^\alpha X_i\|_{\infty,\infty} + \|\partial^\alpha Y_i\|_{\infty,\infty} \quad \leq \quad C, \quad i \in \{1,2\} \quad and$$
$$\|\partial^\beta(X_1 - X_2)\|_{\infty,\infty} + \|\partial^\beta(Y_1 - Y_2)\|_{\infty,\infty} \quad \leq \quad CT(\|\xi_1 - \xi_2\|_\infty + \|\Omega_1 - \Omega_2\|_\infty)$$

for all multi-indices $1 \leq |\alpha| \leq 3$ *and* $0 \leq |\beta| \leq 3$. *Moreover, the constants do not depend on* ξ_i *or* Ω_i, $i \in \{1,2\}$, *but only on their norms* $K_i := \|\xi_i\|_\infty + \|\Omega_i\|_\infty$, $i \in \{1,2\}$.

The remainder of this section is devoted to the proof of this proposition. We split the proof into four lemmas which roughly correspond to steps (2) - (5). It is always assumed that $\xi_1, \xi_2, \Omega_1, \Omega_2 \in W^{1,p}(J_T)$ are given and the index $i \in \{1, 2\}$ on a function means that it is associated to ξ_i, Ω_i by the definitions in steps (1) - (5).

Note that the generic constants which appear in the estimates may depend on the K_i and on T. In Chapter 4, where we use the estimates again, we will choose $0 < T \leq T_0 < \infty$ and ξ_i, Ω_i such that there is a uniform upper bound on K_i for all possible ξ_i, Ω_i so that we may ignore this dependence.

Lemma 2.2. *The vector-valued functions* η_i, ω_i *associated to* ξ_i, Ω_i *in step* (2) *satisfy*

$$\|\eta_1 - \eta_2\|_\infty \leq C(\|\xi_1 - \xi_2\|_\infty + \|\Omega_1 - \Omega_2\|_\infty) \quad and$$
$$\|\omega_1 - \omega_2\|_\infty \leq C\|\Omega_1 - \Omega_2\|_\infty.$$

Proof. First consider the $Q_i \in W^{2,p}(J_T; \mathbb{R}^{3 \times 3})$. From (2.13), we get the estimate

$$|Q_i(t)| \leq C + \int_0^t |M_i(s)|\,|Q_i(s)|\,\mathrm{d}s$$

so that by Gronwall's Inequality,

$$|Q_i(t)| \leq C + \int_0^t C\,|M_i(s)|\,e^{\int_0^s |M_i(\tau)|\,\mathrm{d}\tau}\,\mathrm{d}s.$$

This yields

$$\|Q_i\|_\infty + \|Q_i^T\|_\infty \leq C(1 + T K_i e^{T K_i}).$$

Similarly, we use

$$\dot{Q}_1 - \dot{Q}_2 = M_1 Q_1 - M_2 Q_2, \qquad (Q_1 - Q_2)(0) = 0$$

to get the estimate

$$|(Q_1 - Q_2)(t)| \leq \int_0^t |(M_1 - M_2)(s)|\,|Q_1(s)| + |M_2(s)|\,|(Q_1 - Q_2)(s)|\,\mathrm{d}s$$
$$\leq CT\|M_1 - M_2\|_\infty + \int_0^t |M_2(s)|\,|(Q_1 - Q_2)(s)|\,\mathrm{d}s.$$

Gronwall's Lemma implies

$$\|Q_1 - Q_2\|_\infty \leq CT\|M_1 - M_2\|_\infty \leq CT\|\Omega_1 - \Omega_2\|_\infty. \tag{2.19}$$

Now

$$\|\eta_1 - \eta_2\|_\infty \leq \|Q_1(\xi_1 - \xi_2) + (Q_1 - Q_2)\xi_2\|_\infty \leq C(\|\xi_1 - \xi_2\|_\infty + \|\Omega_1 - \Omega_2\|_\infty)$$

and the estimate

$$\|\omega_1 - \omega_2\|_\infty \leq C \|\Omega_1 - \Omega_2\|_\infty$$

follows analogously.   □

Lemma 2.3. *The maps b_i associated to η_i, ω_i by step (3) satisfy*

$$\|\partial^\beta b_i\|_{\infty,\infty} \leq C \quad and$$
$$\|\partial^\beta(b_1 - b_2)\|_{\infty,\infty} \leq C(\|\omega_1 - \omega_2\|_\infty + \|\eta_1 - \eta_2\|_\infty)$$

for all multi-indices $0 \leq |\beta| \leq 3$.

Proof. First note that by Proposition 1.9, $B_K \in L(W_0^{3,p}(K), W_0^{4,p}(K))$ for $1 < p < \infty$. By Sobolev's embedding theorem we can show that B_K is a bounded operator on $C^3(K)$ using the crude estimate

$$\|B_K f\|_{C^3(K)} \leq C \|B_K f\|_{W^{4,p}(K)} \leq C \|f\|_{W^{3,p}(K)} \leq C \|f\|_{C^3(K)}, \quad p > 3.$$

We now only have to calculate the derivatives of b_i in terms of the derivatives of χ and use Lipschitz continuity of χ and its derivatives to show that

$$\begin{aligned}
\|\partial^\alpha b_i\|_{\infty,\infty} &\leq C \|\chi\|_{C^{|\alpha|+1}(\overline{B_2})} (\|\omega_i\|_\infty + \|\eta_i\|_\infty) \qquad (2.20)\\
&\leq \|Q_i\|_\infty \|\xi_i\|_\infty + \|Q_i\|_\infty \|\Omega_i\|_\infty \leq C_{K_i}
\end{aligned}$$

and also

$$\|\partial^\alpha(b_1 - b_2)\|_{\infty,\infty} \leq C \|\chi\|_{C^{|\alpha|+1}(\overline{B_2})} (\|\omega_1 - \omega_2\|_\infty + \|\eta_1 - \eta_2\|_\infty).$$

□

A similar technique involving Gronwall's Inequality allows us to give estimates for the transforms X_i, Y_i with respect to b_i and $b_i^{(Y)}$.

Lemma 2.4. *The coordinate transforms X_i and Y_i associated to $b_i, b_i^{(Y)}$ by steps (3) and (5), respectively, satisfy*

$$\|\partial^\alpha X_i\|_{\infty,\infty} \leq C$$

for multi-indices $1 \leq |\alpha| \leq 3$ and

$$\|\partial^\beta(X_1 - X_2)\|_{\infty,\infty} \leq CT \|b_1 - b_2\|_{L^\infty(J_T;C^{|\beta|}(\mathbb{R}^3))},$$
$$\|\partial^\beta(Y_1 - Y_2)\|_{\infty,\infty} \leq CT \|b_1^{(Y)} - b_2^{(Y)}\|_{L^\infty(J_T;C^{|\beta|}(\mathbb{R}^3))},$$

for multi-indices $0 \leq |\beta| \leq 3$.

Proof. First we show the second estimate for $|\alpha| = 0$. Our starting point are the differential equations

$$
\begin{cases}
\partial_t (X_1 - X_2)(t, y) = b_1(t, X_1(t, y)) - b_2(t, X_2(t, y)), & t \in J_T, y \in \mathbb{R}^3, \\
(X_1 - X_2)(0, y) = 0, & y \in \mathbb{R}^3.
\end{cases}
$$

By integration in time, Lemma 2.3 and the Lipschitz continuity of b_i, we get

$$
\begin{aligned}
&|(X_1 - X_2)(t, y)| \\
\leq\ & \int_0^t |b_1(s, X_1(s, y)) - b_2(s, X_2(s, y))|\ \mathrm{d}s \\
\leq\ & \int_0^t \|\nabla b_1(s)\|_{L^\infty(\mathbb{R}^3)} |(X_1 - X_2)(s, y)| + \|(b_1 - b_2)(s)\|_{L^\infty(\mathbb{R}^3)}\ \mathrm{d}s
\end{aligned}
$$

for all $(t, y) \in J_T \times \mathbb{R}^3$. This yields

$$
\|X_1 - X_2\|_{\infty,\infty} \leq CT \|b_1 - b_2\|_{\infty,\infty}
$$

by Gronwall's Inequality. Let now $(z_i)_{kj}(t, y) := \partial_j (X_i)_k(t, y)$, $j, k \in \{1, 2, 3\}$ the partial derivative of the k-th component of X_i with respect to its j-th argument and let J_{b_i} the Jacobian of b_i. By differentiating (2.8) with respect to the spatial variables we get

$$
\begin{cases}
\partial_t (z_i)_{kj}(t, y) = J_{b_i}(t, X_i(t, y))(z_i)_{kj}(t, y), & t \in J_T, y \in \mathbb{R}^3, \\
(z_i)_{kj}(0, y) = e_j, & y \in \mathbb{R}^3,
\end{cases}
$$

where e_j denotes the j-th unit vector in $\mathbb{R}^3$. So Gronwall's Inequality and Lemma 2.3 imply

$$
\|(z_i)_{kj}\|_{\infty,\infty} \leq e^{T\|J_{b_i}\|_{\infty,\infty}} \leq C.
$$

For two different transforms, the relation

$$
\begin{cases}
\partial_t ((z_1)_{kj} - (z_2)_{kj})(t, y) = J_{b_1}(t, X_1(t, y))(z_1)_{kj}(t, y) \\
\qquad\qquad\qquad - J_{b_2}(t, X_2(t, y))(z_2)_{kj}(t, y), & t \in J_T, y \in \mathbb{R}^3, \\
((z_1)_{kj} - (z_2)_{kj})(0, y) = 0, & y \in \mathbb{R}^3.
\end{cases}
$$

holds true. Again, we integrate to show

$$
\begin{aligned}
|((z_1)_{kj} - (z_2)_{kj})(t, y)| \leq\ & T\big(\|\nabla(b_1 - b_2)\|_{\infty,\infty} \\
& + \|D^2 b_2\|_{\infty,\infty} \|X_1 - X_2\|_{\infty,\infty}\big) \|(z_1)_{kj}\|_{\infty,\infty} \\
& + \int_0^t |J_{b_2}(s, X_2(s, y))| |((z_1)_{kj} - (z_2)_{kj})(s, y)|\ \mathrm{d}s.
\end{aligned}
$$

Gronwall's Inequality and the estimates obtained above yield

$$\|(z_1)_{kj} - (z_2)_{kj}\|_{\infty,\infty} \leq CT \|\nabla(b_1 - b_2)\|_{\infty,\infty} + CT^2 \|b_1 - b_2\|_{\infty,\infty} . \quad (2.21)$$

The second and third derivatives can be done in a similar way, if we consider the equations

$$\begin{cases} \partial_t(z_i)_{kjl}(t,y) = \partial_k J_{b_i}(t, X_i(t,y))(z_i)_{kjl}(t,y) \\ \qquad\qquad + J_{b_i}(t, X_i(t,y))(z_i)_{kjl}(t,y), & t \in J_T, y \in \mathbb{R}^3, \\ (z_i)_{kjl}(0,y) = 0, & y \in \mathbb{R}^3 \end{cases}$$

for $(z_i)_{kjl}(t,y) := \partial_j \partial_k (X_i)_l(t,y)$ and

$$\begin{cases} \partial_t z_{ijkl}(t,y) = \partial_l \partial_k J_{b_i}(t, X_i(t,y))(z_i)_{kj}(t,y) \\ \qquad\qquad + \partial_k J_{b_i}(t, X_i(t,y))(z_i)_{kjl}(t,y) \\ \qquad\qquad + \partial_l J_{b_i}(t, X_i(t,y))(z_i)_{kjl}(t,y) \\ \qquad\qquad + J_{b_i}(t, X_i(t,y))(z_i)_{kjlm}(t,y), & t \in J_T, y \in \mathbb{R}^3, \\ (z_i)_{kjlm}(0,y) = 0, & y \in \mathbb{R}^3 \end{cases}$$

for $(z_i)_{kjlm}(t,y) := \partial_j \partial_k \partial_l (X_i)_m(t,y)$. Clearly, the same arguments yield the estimates for Y. $\qquad\qquad\square$

Since the $b_i^{(Y)}$ are defined implicitly through the transforms X_i, it remains to show that this does not worsen its dependence on the ξ_i and Ω_i considerably.

Lemma 2.5. *Let b_i be given by η_i, ω_i as in step (3) and $b_i^{(Y)}$ correspondingly as in step (5), then*

$$\|b_1^{(Y)} - b_2^{(Y)}\|_{L^\infty(J_T; C^3(\mathbb{R}^3))} \leq C \|b_1 - b_2\|_{L^\infty(J_T; C^4(\mathbb{R}^3))} .$$

Proof. A simple calculation shows that for all $(t,y) \in [0,T) \times \mathbb{R}^3$,

$$\begin{aligned} \|b_1^{(Y)} - b_2^{(Y)}\|_{\infty,\infty} &\leq \|J_{X_1}^{-1} - J_{X_2}^{-1}\|_{\infty,\infty} \|b_1\|_{\infty,\infty} \\ &\quad + \|J_{X_2}^{-1}\|_{\infty,\infty} \|\nabla b_1\|_{\infty,\infty} \|X_1 - X_2\|_{\infty,\infty} \\ &\quad + \|J_{X_2}^{-1}\|_{\infty,\infty} \|b_1 - b_2\|_{\infty,\infty} , \end{aligned}$$

where

$$\|J_{X_1}^{-1} - J_{X_2}^{-1}\|_{\infty,\infty} \leq \|J_{X_1}^{-1}\|_{\infty,\infty} \|J_{X_2} - J_{X_1}\|_{\infty,\infty} \|J_{X_2}^{-1}\|_{\infty,\infty}.$$

By Lemma 2.4, we get the desired estimate if we can show that $\|J_{X_i}^{-1}\|_{\infty,\infty} \leq C$. This is obvious for fixed ξ_i, Ω_i since the inversion of a matrix is continuous in its entries and $J_{X_i}(t,y) = \mathrm{Id}_{\mathbb{R}^3}$ for all $t \in [0,T]$ and y in the complement of $K \cup B_2$. However, we need additional arguments to show that C can really be chosen independently of ξ_i and Ω_i themselves and only involves there sup-norms K_i. Since $\|J_{X_i}\|_{\infty,\infty} \leq C_{K_i}$ uniformly by Lemma 2.4, the $1 \leq k \leq 3$ eigenvalues $\lambda_{ik}(t,y)$ of multiplicity $1 \leq s_{ik}(t,y) \leq 3$ of $J_{X_i}(t,y)$ satisfy $\|\lambda_{ik}\|_{\infty,\infty} \leq C_{K_i}$, so the inverse eigenvalues λ_{ik}^{-1} stay away from zero, $\|\frac{1}{\lambda_{ik}}\|_{\infty,\infty} \geq \frac{1}{C_{K_i}}$, uniformly in t and y. Since $\det J_{X_i}^{-1} = 1$ by (2.9), we get that $1 = \Pi_k(\frac{1}{\lambda_{ik}(t,y)})^{s_{ik}(t,y)}$ and thus $\max_k(\|\frac{1}{\lambda_{ik}}\|_{\infty,\infty}) \leq C_{K_i}^2$. The uniform upper bound on the eigenvalues also yields the boundedness of $J_{X_i}^{-1}$.

In a similar way, we can treat the derivatives of the $b_i^{(Y)}$. The first partial derivatives satisfy

$$
\begin{aligned}
&\|\partial_j(b_1^{(Y)} - b_2^{(Y)})\|_{\infty,\infty} \\
\leq\ & C\big[\|\partial_j(J_{X_1}^{-1} - J_{X_2}^{-1})\|_{\infty,\infty} + \|\partial_j J_{X_2}^{-1}\|_{\infty,\infty}\|b_1(\cdot,X_1) - b_2(\cdot,X_2)\|_{\infty,\infty} \\
&+ \|J_{X_1}^{-1} - J_{X_2}^{-1}\|_{\infty,\infty}\|\partial_j X_1\|_{\infty,\infty} \\
&+ \|J_{X_2}^{-1}\|_{\infty}(\|J_{b_1}(\cdot,X_1) - J_{b_2}(\cdot,X_2)\|_{\infty,\infty}\|\partial_j X_1\|_{\infty,\infty} \\
&+ \|\partial_j(X_1 - X_2)\|_{\infty,\infty})\big] \\
\leq\ & C\big[\|\partial_j(J_{X_1}^{-1}(J_{X_2} - J_{X_1})J_{X_2}^{-1})\|_{\infty,\infty} + \|b_1(\cdot,X_1) - b_1(\cdot,X_2)\|_{\infty,\infty} \\
&+ \|b_1(\cdot,X_2) - b_2(\cdot,X_2)\|_{\infty,\infty} + \|b_1 - b_2\|_{L^\infty(J_T;C^1(\mathbb{R}^3))} \\
&+ \|J_{b_1}(\cdot,X_1) - J_{b_1}(\cdot,X_2)\|_{\infty,\infty} + \|J_{b_1}(\cdot,X_2) - J_{b_2}(\cdot,X_2)\|_{\infty,\infty}\big] \\
\leq\ & C\big[\|J_{X_2} - J_{X_1}\|_{\infty,\infty} + \|\partial_j(J_{X_2} - J_{X_1})\|_{\infty,\infty} \\
&+ (\|\nabla b_1\|_{\infty,\infty} + \|\nabla^2 b_1\|_{\infty,\infty})\|X_1 - X_2\|_{\infty,\infty} + \|b_1 - b_2\|_{L^\infty(J_T;C^1(\mathbb{R}^3))}\big] \\
\leq\ & C\|b_1 - b_2\|_{L^\infty(J_T;C^2(\mathbb{R}^3))} ,
\end{aligned}
$$

by Lemma 2.1 and the estimates obtained above. The proof of the estimate for the second partial derivatives of $b_1^{(Y)} - b_2^{(Y)}$ can be done analogously. $\quad\sqcup$

It is now easy to prove Proposition 2.1 by putting the above lemmas together. The first estimate on $\partial^\alpha X_i$ was proved already in Lemma 2.4. The boundedness of the $\partial^\alpha Y_i$ follows from the arguments used for $J_{X_i}^{-1}$ in the proof of Lemma 2.5. Furthermore, the differences satisfy

$$
\begin{aligned}
\left\|\partial^\beta(X_1 - X_2)\right\|_{\infty,\infty} &\leq CT\|b_1 - b_2\|_{L^\infty(J_T;C^{|\beta|}(\mathbb{R}^3))} \\
&\leq CT(\|\eta_1 - \eta_2\|_\infty + \|\omega_1 - \omega_2\|_\infty) \\
&\leq CT(\|\xi_1 - \xi_2\|_\infty + \|\Omega_1 - \Omega_2\|_\infty)
\end{aligned}
$$

by Lemmas 2.4, 2.3 and 2.2. Adding Lemma 2.5 similarly yields

$$
\begin{aligned}
\left\|\partial^{\beta}(Y_1 - Y_2)\right\|_{\infty,\infty} &\leq CT \|b_1^{(Y)} - b_2^{(Y)}\|_{L^{\infty}(J_T; C^{|\beta|}(\mathbb{R}^3))} \\
&\leq CT \|b_1 - b_2\|_{L^{\infty}(J_T; C^{|\beta|+1}(\mathbb{R}^3))} \\
&\leq CT(\|\xi_1 - \xi_2\|_{\infty} + \|\Omega_1 - \Omega_2\|_{\infty}).
\end{aligned}
$$

Chapter 3

Maximal Regularity of the Linear System

The aim of this and the following chapter is to show the existence of a unique local strong solution to the transformed problem (2.17). We linearize the fluid part by the classical Stokes equations with inhomogeneous boundary conditions and we linearize the rigid body equations using the original stress tensor $\mathbf{T}(u, p)$ instead of $\mathcal{T}(u, p)$. We show that the remaining coupled system has maximal regularity.

In a second step, the leftover non-linearities can be treated via a contraction mapping argument on small time intervals. This is done in Chapter 4.

We consider the linear system

$$
\begin{cases}
\begin{aligned}
u_t - \Delta u + \nabla p &= f_0 & &\text{in } J_{T_0} \times \mathcal{D}, \\
\operatorname{div} u &= 0 & &\text{in } J_{T_0} \times \mathcal{D}, \\
u(t, y) &= \xi(t) + \Omega(t) \times y, & &t \in J_{T_0}, y \in \Gamma, \\
u(0) &= u_0 & &\text{in } \mathcal{D}, \\
\mathrm{m}\xi' + \int_\Gamma \mathbf{T}(u, p) N \, \mathrm{d}\sigma &= f_1 & &\text{in } J_{T_0}, \\
I\Omega' + \int_\Gamma y \times \mathbf{T}(u, p) N \, \mathrm{d}\sigma &= f_2 & &\text{in } J_{T_0}, \\
\xi(0) &= \xi_0, \\
\Omega(0) &= \Omega_0,
\end{aligned}
\end{cases}
\tag{3.1}
$$

for some $T_0 > 0$. This is motivated by the following procedure. We first add $(\mathcal{L} - \Delta)u$ and $(\nabla - \mathcal{G})p$, defined in Section 2.2, to the first line of (2.17) and $\int_\Gamma \mathbf{T}(u, p) N \, \mathrm{d}\sigma$, $\int_\Gamma y \times \mathbf{T}(u, p) N \, \mathrm{d}\sigma$ to the equations for the rigid body. Then we move all non-linear terms to the right hand side and pretend they are fixed. Here, we consider the terms $\mathcal{M}u, \mathcal{L}u, \ldots$ as non-linear, because the operators $\mathcal{M}$ and $\mathcal{L}$ depend on the change of coordinates and therefore on the body velocity.

Our aim is to show that for any given $f_0 \in L^p(J_{T_0}; L^q(\mathcal{D}; \mathbb{R}^3))$, $f_1, f_2 \in L^p(J_{T_0}; \mathbb{R}^3)$ and suitable initial data there is a strong solution

$$
\begin{aligned}
u &\in X_{p,q}^{T_0} = W^{1,p}(J_{T_0}; L^q(\mathcal{D})) \cap L^p(J_{T_0}; W^{2,q}(\mathcal{D})) \\
p &\in Y_{p,q}^{T_0} = L^p(J_{T_0}; \widehat{W}^{1,q}(\mathcal{D})), \\
(\xi, \Omega) &\in W^{1,p}(J_{T_0}; \mathbb{R}^6)
\end{aligned}
$$

of (3.1). More precisely, we require that $\xi_0, \Omega_0 \in \mathbb{R}^3$ and that $u_0 \in B_{q,p}^{2-2/p}(\mathcal{D})$ satisfies the compatibility conditions

$$\operatorname{div} u_0 = 0 \tag{3.2}$$

and

$$
\begin{aligned}
u_0(y) &= \xi_0 + \Omega_0 \times y, & y \in \Gamma, \text{ if } \tfrac{1}{2q} + \tfrac{1}{p} < 1 \text{ or} \\
u_0(y) \cdot N(y) &= (\xi_0 + \Omega_0 \times y) \cdot N(y), & y \in \Gamma, \text{ if } \tfrac{1}{2q} + \tfrac{1}{p} > 1.
\end{aligned} \tag{3.3}
$$

They are a consequence of the way we solve the reference problem, see (3.4) below, and of the characterization of the time-trace space $(L_\sigma^q, D(A_q))_{1-1/p,p}$ of the Stokes problem in Subsection 1.3.1.

The main result of this chapter is the following theorem.

Theorem 3.1. *Let $p, q \in (1, \infty)$ such that $\frac{3}{2q} + \frac{1}{p} \leq \frac{3}{2}$, $\frac{1}{2q} + \frac{1}{p} \neq 1$, let $T_0 > 0$ and let $\mathcal{D}$ be an exterior domain with boundary of class $C^{2,1}$. Assume that ξ_0, Ω_0, u_0 and f_0, f_1, f_2 satisfy the above conditions. Then problem (3.1) admits a unique solution*

$$u \in X_{p,q}^{T_0}, \quad p \in Y_{p,q}^{T_0}, \quad (\xi, \Omega) \in W^{1,p}(J_{T_0}; \mathbb{R}^6),$$

which satisfies the estimate

$$
\begin{aligned}
&\|u\|_{X_{p,q}^{T_0}} + \|p\|_{Y_{p,q}^{T_0}} + \|(\xi, \Omega)\|_{W^{1,p}(J_{T_0})} \\
&\leq C(\|f_0\|_{p,q} + \|(f_1, f_2)\|_p + \|u_0\|_{B_{q,p}^{2-2/p}(\mathcal{D})} + |(\xi_0, \Omega_0)|),
\end{aligned}
$$

where the constant C depends only on the geometry of the body and on T_0.

The remainder of this chapter is devoted to the proof of this theorem. We will make extensive use of the preliminary results on the Stokes problem given in Sections 1.3-1.6. In Section 3.1, we give a reformulation of (3.1) as a linear fixed point equation in the unknowns ξ, Ω only, which is necessary in order to deal with the strong coupling between the fluid and the body parts. In Section 3.2, the reformulated problem is solved and it is explained how to retrieve Theorem 3.1. In this context, note the special structure of the coupling. Given u, p, the body velocities ξ and Ω can be obtained by integrating lines 5 and 6 of (3.1) in time. On the other hand, given ξ, Ω, we get solutions u, p of the fluid part of the linear problem which depend continuously on the data in $W^{1,p}(J_{T_0})$. If we plug this solution operator into lines 5 and 6 and thereby eliminate the unknowns u, p, this information does not yield an ordinary differential equation in $\mathbb{R}^6$ but only a linear equation in $W^{1,p}(J_{T_0})$, cf. (3.15). A second difficulty of this system is that it is not clear how to localize the problem in the spatial coordinates. A formulation of the fluid equations in the half-space would make the body infinitely large, so that the forces, e.g. $\int_\Gamma \mathbf{T}(u, p) N \, d\sigma$ become unbounded. Therefore it seems easier to try to go back to known results on the Stokes problem on exterior domains.

3.1   A Reformulation of the Problem

The reformulation procedure can be split into three smaller steps. First we obtain homogeneous initial conditions for problem (3.1) by subtracting the solution $u^* = \mathcal{U}(f_0, \xi_0 + \Omega_0 \times y, u_0)$ and $p^* = \mathcal{P}(f_0, \xi_0 + \Omega_0 \times y, u_0)$ of

$$\begin{cases} u_t^* - \Delta u^* + \nabla p^* &= f_0 & \text{in } J_{T_0} \times \mathcal{D}, \\ \operatorname{div} u^* &= 0 & \text{in } J_{T_0} \times \mathcal{D}, \\ u^*(t, y) &= \xi_0 + \Omega_0 \times y, & t \in J_{T_0}, y \in \Gamma, \\ u^*(0) &= u_0 & \text{in } \mathcal{D}, \end{cases} \tag{3.4}$$

from u, p and ξ_0, Ω_0 from ξ, Ω, respectively. Note that the estimate

$$\|u^*\|_{X_{p,q}^{T_0}} + \|p^*\|_{Y_{p,q}^{T_0}} \le C(\|u_0\|_{B_{q,p}^{2-2/p}(\mathcal{D})} + \|f_0\|_{p,q} + |(\xi_0, \Omega_0)|), \tag{3.5}$$

follows from (1.19) and if we estimate the auxiliary function b_h from Section 1.5 as in (2.20).

The next step is a modification of the Dirichlet interface condition. Even though the system (3.1) arises naturally as a linearization of the transformed equations, it is not satisfactory from a physical point of view, as the interface condition

$$u|_\Gamma(t, y) = \xi(t) + \Omega(t) \times y$$

does not satisfy $u|_\Gamma \cdot N = 0$ in general and therefore suggests that the fluid is allowed to flow out of $\mathcal{D}$. As a remedy, we define a new fluid velocity $\hat{u}$ which satisfies a Stokes problem with boundary condition $\hat{u}|_\Gamma \cdot N = 0$ and absorb the normal component of the body velocity into a modified pressure $\hat{p}$. In the rigid body equations, this additional part of the pressure acts as an added mass.

Let now $e_1 = (1, 0, 0), e_2 = (0, 1, 0), e_3 = (0, 0, 1)$ and let $v^{(i)}, V^{(i)}$ be solutions of the weak Neumann problems

$$
\begin{cases}
\Delta v^{(i)} &= 0, & \text{in } \mathcal{D}, \\
\frac{\partial v^{(i)}}{\partial N}|_\Gamma &= N \cdot e_i, & \text{on } \Gamma,
\end{cases}
$$

$$
\begin{cases}
\Delta V^{(i)} &= 0, & \text{in } \mathcal{D}, \\
\frac{\partial V^{(i)}}{\partial N}|_\Gamma(y) &= N(y) \cdot (e_i \times y), & y \in \Gamma,
\end{cases}
$$

where $N(y)$ denotes the outer normal of $\mathcal{B}$. By [GHHS], we obtain strong regularity and

$$
v^{(i)}, V^{(i)} \in \bigcap_{1 \leq m \leq 3} \widehat{W}^{m,r}(\mathcal{D}) \tag{3.6}
$$

for every $1 < r < \infty$. We now have a basis for our correction. Let $0 < T \leq T_0$. For any given $\xi, \Omega \in {}_0W^{1,p}(J_T)$, we define

$$
v_{\xi,\Omega}(t) := \sum_i \xi_i(t) v^{(i)} + \Omega_i(t) V^{(i)} \qquad \text{for all } t \in J_T, \tag{3.7}
$$

which implies that

$$
\begin{cases}
\Delta v_{\xi,\Omega}(t) &= 0, & \text{in } \mathcal{D}, \\
\frac{\partial v_{\xi,\Omega}(t)}{\partial N}|_\Gamma &= (\xi(t) + \Omega(t) \times y) \cdot N, & \text{on } \Gamma,
\end{cases}
$$

is satisfied. It follows immediately from the definition that

$$
\|\nabla v_{\xi,\Omega}\|_{W^{1,p}(J_T;W^{2,q}(\mathcal{D}))} + \|\partial_t v_{\xi,\Omega}\|_{Y_{p,q}^T} \leq C \|(\xi, \Omega)\|_{W^{1,p}(J_T)}.
$$

In the following, we define new unknown functions $\hat{u}, \hat{p}, \hat{\xi}, \hat{\Omega}$ by

$$
\begin{aligned}
u &:= u^* + \hat{u} + \nabla v_{\xi,\Omega}, \\
p &:= p^* + \hat{p} - \partial_t v_{\xi,\Omega}, \\
\xi &:= \xi_0 + \hat{\xi}, \\
\Omega &:= \Omega_0 + \hat{\Omega}.
\end{aligned}
$$

Instead of (3.1), we consider the equivalent problem

$$
\begin{cases}
\hat{u}_t - \Delta\hat{u} + \nabla\hat{p} = 0 & \text{in } J_T \times \mathcal{D}, \\
\operatorname{div}\hat{u} = 0 & \text{in } J_T \times \mathcal{D}, \\
\hat{u}|_\Gamma = h(\hat{\xi}, \hat{\Omega}) & \text{on } J_T \times \Gamma, \\
\hat{u}(0) = 0 & \text{in } \mathcal{D}, \\
m\hat{\xi}' + \int_\Gamma \partial_t v_{\hat{\xi}, \hat{\Omega}} N \, d\sigma = f_1 - \int_\Gamma \mathbf{T}(u, p)N \, d\sigma & \text{in } J_T, \\
I\hat{\Omega}' + \int_\Gamma y \times (\partial_t v_{\hat{\xi}, \hat{\Omega}} N) \, d\sigma = f_2 - \int_\Gamma y \times \mathbf{T}(u, p)N \, d\sigma & \text{in } J_T, \\
\hat{\xi}(0) = 0, \\
\hat{\Omega}(0) = 0,
\end{cases}
\tag{3.8}
$$

where

$$
h(\hat{\xi}, \hat{\Omega}) := \hat{\xi} + \hat{\Omega} \times y - \nabla v_{\hat{\xi}, \hat{\Omega}}|_\Gamma.
\tag{3.9}
$$

Due to the correction by $v_{\hat{\xi}, \hat{\Omega}}$, we get the additional condition $\hat{u}|_\Gamma N = 0$ on the boundary. As $v_{\hat{\xi}, \hat{\Omega}}$ was defined to absorb the normal component of the interface velocity into the pressure, the correction $\nabla v_{\hat{\xi}, \hat{\Omega}}$ applied to u does not affect the rigid body. This can be seen from the following calculations. It holds that

$$
\begin{aligned}
\left(\int_\Gamma \mathcal{E}^{(\nabla v_{\hat{\xi}, \hat{\Omega}})} N \, d\sigma \right)_i &= \int_\Gamma (\partial_i \nabla v_{\hat{\xi}, \hat{\Omega}}) \cdot N \, d\sigma \\
&= \int_\mathcal{D} \operatorname{div}(\partial_i \nabla v_{\hat{\xi}, \hat{\Omega}}) = \int_\mathcal{D} \partial_i \Delta v_{\hat{\xi}, \hat{\Omega}} = 0.
\end{aligned}
$$

Similarly, we calculate the components of $\int_\Gamma y \times \mathcal{E}^{(\nabla v_{\hat{\xi}, \hat{\Omega}})} N \, d\sigma$ to show that $\mathcal{J}\mathcal{E}^{(\nabla v_{\hat{\xi}, \hat{\Omega}})} = 0$. Let $i, j, k \in \{1, 2, 3\}$ and $i \neq j \neq k$ such that we get the correct sign, then

$$
\begin{aligned}
\left(\int_\Gamma y \times \mathcal{E}^{(\nabla v_{\hat{\xi}, \hat{\Omega}})} N \, d\sigma \right)_i &= \int_\mathcal{D} \operatorname{div}\left((-y_j \partial_k + y_k \partial_j) \nabla v_{\hat{\xi}, \hat{\Omega}} \right) \\
&= \int_\mathcal{D} (\partial_j \partial_k - \partial_k \partial_j) v_{\hat{\xi}, \hat{\Omega}} + (-y_j \partial_k + y_k \partial_j) \Delta v_{\hat{\xi}, \hat{\Omega}} \\
&= 0.
\end{aligned}
$$

In both cases, we can show by a density argument that the Gauss Theorem can be applied as the integrals exist.

For convenience of notation and in order to reformulate the problem, let

$$
\mathbb{I} := \begin{pmatrix} m\mathrm{Id}_{\mathbb{R}^3} & 0 \\ 0 & I \end{pmatrix}
\tag{3.10}
$$

the constant momentum matrix for the problem and let $\mathcal{J} : (W_{loc}^{1,q}(\mathcal{D}))^{3\times 3} \to \mathbb{R}^6$ the integral operator given by

$$\mathcal{J}h = \begin{pmatrix} \int_\Gamma hN \, d\sigma \\ \int_\Gamma y \times hN \, d\sigma \end{pmatrix}. \tag{3.11}$$

Furthermore, we define an *added mass* $\mathbb{M}$ of the body in the following way, cf. [Gal02, p. 685]. Let $a_{ij}, b_{ij}, c_{ij}, d_{ij}$ be given by

$$a_{ij} := \int_\Gamma v^{(i)} N_j \, d\sigma,$$

$$b_{ij} := \int_\Gamma V^{(i)}(e_j \times y) \cdot N \, d\sigma,$$

$$c_{ij} := \int_\Gamma v^{(i)}(e_j \times y) \cdot N \, d\sigma,$$

$$d_{ij} := \int_\Gamma V^{(i)} N_j \, d\sigma$$

and let

$$\mathbb{M} := \begin{pmatrix} a_{11} & a_{12} & a_{13} & c_{11} & c_{12} & c_{13} \\ a_{21} & a_{22} & a_{23} & c_{21} & c_{22} & c_{23} \\ a_{31} & a_{32} & a_{33} & c_{31} & c_{32} & c_{33} \\ d_{11} & d_{12} & d_{13} & b_{11} & b_{12} & b_{13} \\ d_{21} & d_{22} & d_{23} & b_{21} & b_{22} & b_{23} \\ d_{31} & d_{32} & d_{33} & b_{31} & b_{32} & b_{33} \end{pmatrix}. \tag{3.12}$$

From this definition, we get that $\mathcal{J}(\partial_t v_{\hat{\xi},\hat{\Omega}}) = \mathbb{M} \begin{pmatrix} \hat{\xi}' \\ \hat{\Omega}' \end{pmatrix}$, so that the body equations in (3.8) can be rewritten as

$$(\mathbb{I} + \mathbb{M}) \begin{pmatrix} \hat{\xi}' \\ \hat{\Omega}' \end{pmatrix} = \begin{pmatrix} f_1 \\ f_2 \end{pmatrix} - \mathcal{J}\mathbf{T}(\hat{u}, \hat{p}) - \mathcal{J}\mathbf{T}(u^*, p^*).$$

Furthermore, we obtain the following properties of the added mass matrix.

Lemma 3.2. *The matrix $\mathbb{M}$ is symmetric and semi positive-definite.*

Proof. The matrix $\mathbb{M}$ is symmetric because by the Gauss theorem and the properties of $v^{(i)}, V^{(i)}$,

$$a_{ij} = \int_\Gamma v^{(i)} N_j \, d\sigma = \int_\Gamma v^{(i)} \frac{\partial v^{(j)}}{\partial N} \, d\sigma = \int_\mathcal{D} \nabla v^{(i)} \cdot \nabla v^{(j)} = \sum_{l=1}^3 \int_\mathcal{D} \partial_l v^{(i)} \partial_l v^{(j)},$$

and similarly,

$$b_{ij} = \sum_{l=1}^{3} \int_{\mathcal{D}} \partial_l V^{(i)} \partial_l V^{(j)} = b_{ji} \quad \text{and} \quad c_{ij} = \sum_{l=1}^{3} \int_{\mathcal{D}} \partial_l v^{(i)} \partial_l V^{(j)} = d_{ji}.$$

The existence of these integrals follows from (3.6) for $r = 2$. Now consider any vector $z = (x_1, x_2, x_3, y_1, y_2, y_3) \in \mathbb{R}^6$, then

$$
\begin{aligned}
z^T \mathbb{M} z &= \sum_{i,j=1}^{3} a_{ij} x_i x_j + \sum_{i,j=1}^{3} c_{ij} x_i y_j + \sum_{i,j=1}^{3} d_{ij} y_i x_j + \sum_{i,j=1}^{3} b_{ij} y_i y_j \\
&= \sum_{l=1}^{3} \int_{\mathcal{D}} \Big(\sum_{i,j=1}^{3} \partial_l v^{(i)} \partial_l v^{(j)} x_i x_j + \sum_{i,j=1}^{3} \partial_l v^{(i)} \partial_l V^{(j)} x_i y_j \\
&\quad + \sum_{i,j=1}^{3} \partial_l V^{(i)} \partial_l v^{(j)} y_i x_j + \sum_{i,j=1}^{3} \partial_l V^{(i)} \partial_l V^{(j)} y_i y_j \Big) \\
&= \sum_{l=1}^{3} \Big(\sum_{i=1}^{3} \partial_l v^{(i)} x_i + \Sigma_{i=1}^{3} \partial_l V^{(i)} y_i \Big)^2 \geq 0.
\end{aligned}
$$

$\square$

For every choice of the body's density and mass $\rho_B, \mathrm{m} > 0$, $\mathbb{I}$ is strictly positive, so Lemma 3.2 yields that $\mathbb{I} + \mathbb{M}$ is invertible.

As the final step in this section, we show that the system (3.8) can be rewritten as a problem in the unknowns $\hat{\xi}, \hat{\Omega}$ by using the solution operators $\mathcal{U}, \mathcal{P}$ for the Stokes problem defined in (1.20). Given $\hat{\xi}, \hat{\Omega}$, the fluid part of system (3.8) can be solved by

$$\mathcal{U}_h(\hat{\xi}, \hat{\Omega}) := \mathcal{U}(0, h(\hat{\xi}, \hat{\Omega}), 0) \in X_{p,q,0}^T \qquad (3.13)$$

and

$$\mathcal{P}_h(\hat{\xi}, \hat{\Omega}) := \mathcal{P}(0, h(\hat{\xi}, \hat{\Omega}), 0) \in Y_{p,q}^T. \qquad (3.14)$$

This solution depends linearly and continuously on $\hat{\xi}, \hat{\Omega} \in {}_0W^{1,p}(J_T)$.

Thus, instead of problem (3.8) we can write the equation

$$\begin{pmatrix} \hat{\xi} \\ \hat{\Omega} \end{pmatrix} = \mathcal{R} \begin{pmatrix} \hat{\xi} \\ \hat{\Omega} \end{pmatrix} + f^*, \qquad (3.15)$$

where $\mathcal{R} : {}_0W^{1,p}(J_T; \mathbb{R}^6) \to {}_0W^{1,p}(J_T; \mathbb{R}^6)$ is given by

$$\mathcal{R}(\hat{\xi}, \hat{\Omega})(t) := \int_0^t (\mathbb{I} + \mathbb{M})^{-1} \mathcal{J} \left[\mathbf{T}(\mathcal{U}_h(\hat{\xi}, \hat{\Omega}), \mathcal{P}_h(\hat{\xi}, \hat{\Omega})) \right](s) \, \mathrm{d}s$$

and

$$f^*(t) := \int_0^t (\mathbb{I} + \mathbb{M})^{-1} \left[\begin{pmatrix} f_1 \\ f_2 \end{pmatrix} - \mathcal{J}\mathbf{T}(u^*, p^*) \right](s) \, \mathrm{d}s.$$

In the next section, we show that for sufficiently small $0 < T \leq T_0$ there exists a unique $(\hat{\xi}, \hat{\Omega}) \in {}_0W^{1,p}(J_T; \mathbb{R}^6)$ such that

$$\begin{pmatrix} \hat{\xi} \\ \hat{\Omega} \end{pmatrix} = (\mathrm{Id} - \mathcal{R})^{-1} f^*. \tag{3.16}$$

This is a consequence of the properties of $\mathcal{U}, \mathcal{P}$, the improved time regularity of $\mathcal{P}_h(\hat{\xi}, \hat{\Omega}) \in Y_{p,q}^T$ which follows from Lemma 1.10 and basic estimates for the trace integral operator $\mathcal{J}$.

3.2 Proof of Theorem 3.1

The next lemma yields that $\mathrm{Id} - \mathcal{R}$ is invertible on ${}_0W^{1,p}(J_T; \mathbb{R}^6)$ if the time T is sufficiently small.

Lemma 3.3. *The map $\mathcal{R}$ is bounded linear and $\|\mathcal{R}\|_{\mathcal{L}({}_0W^{1,p}(J_T;\mathbb{R}^6))} < 1$ for sufficiently small $0 < T \leq T_0$. Moreover, $f^* \in {}_0W^{1,p}(J_T; \mathbb{R}^6)$.*

Proof. Let $(\hat{\xi}, \hat{\Omega}) \in {}_0W^{1,p}(J_T; \mathbb{R}^6)$. The functions $H = \hat{\xi} + \hat{\Omega} \times \cdot - \nabla v_{\hat{\xi}, \hat{\Omega}}$ and $h(\hat{\xi}, \hat{\Omega}) = H|_\Gamma$ satisfy the conditions (1.14), so that we get

$$\|\mathcal{U}_h(\hat{\xi}, \hat{\Omega})\|_{X_{p,q}^T} + \|\mathcal{P}_h(\hat{\xi}, \hat{\Omega})\|_{Y_{p,q}^T} \leq C\|(\hat{\xi}, \hat{\Omega})\|_{W^{1,p}(J_T)}$$

by (1.19). Moreover, we can apply Lemma 1.10 to show

$$\|\mathcal{P}_h(\hat{\xi}, \hat{\Omega})\|_{L^p(J_T; L_0^q(\mathcal{D}_R))} \leq CT^{\alpha/p}\|b_{h(\hat{\xi}, \hat{\Omega})}\|_{X_{p,q}^T} \leq CT^{\alpha/p}\|(\hat{\xi}, \hat{\Omega})\|_{W^{1,p}(J_T)} \tag{3.17}$$

for a suitable choice of $R > 0$ and $0 \leq \alpha < \frac{1}{2q}$. Here, $b_{h(\hat{\xi}, \hat{\Omega})} = \chi h(\hat{\xi}, \hat{\Omega}) - B_K(\nabla\chi h(\hat{\xi}, \hat{\Omega})) \in W^{1,p}(J_T; C_{c,\sigma}^\infty(\mathbb{R}^3))$ is the auxiliary function from (1.16), which moves the boundary condition $h(\hat{\xi}, \hat{\Omega})$ to the right hand side of the Stokes equation. By construction, it is solenoidal and it satisfies

$$\partial_t b_{h(\hat{\xi}, \hat{\Omega})}|_\Gamma \cdot N = \partial_t h(\hat{\xi}, \hat{\Omega}) \cdot N = 0$$

and $\Delta b|_\Gamma = 0$ on the boundary, so that the right hand side $\Delta b_{h(\hat{\xi},\hat{\Omega})} - \partial_t b_{h(\hat{\xi},\hat{\Omega})} \in L^p(J_T; L^q_\sigma(\mathcal{D}))$ satisfies the assumption in Lemma 1.10.

For all $0 < \varepsilon < 1 - 1/q$, Proposition 1.1 on the traces of functions $g \in W^{\varepsilon+1/q,q}(\mathcal{D})$ implies that

$$|\mathcal{J}(g)| \leq C \|\gamma g\|_{L^1(\Gamma)} \leq C \|\gamma g\|_{L^q(\Gamma)} \leq C \|g\|_{W^{\varepsilon+1/q,q}(\mathcal{D})}. \tag{3.18}$$

Now choose $\varepsilon = \frac{1}{2} - \frac{1}{2q}$, so that $1 + \frac{1}{q} + \varepsilon = 2 - \varepsilon$. Then Proposition 1.8 yields the embedding

$$X^T_{p,q} \hookrightarrow W^{\varepsilon/2,p}(J_T; W^{2-\varepsilon,q}(\mathcal{D}))$$

and we can use the condition $\frac{3}{2q} + \frac{1}{p} \leq \frac{3}{2}$ to show

$$\frac{\varepsilon}{2} - \frac{3}{p} = \frac{1}{6}\left(\frac{3}{2} - \frac{3}{2q}\right) - \frac{3}{p} \geq -\frac{3 \cdot 17}{18p},$$

so that by Sobolev's embedding theorem,

$$W^{\varepsilon/2,p}(J_T; W^{2-\varepsilon,q}(\mathcal{D})) \hookrightarrow L^{\frac{18}{17}p}(J_T; W^{2-\varepsilon,q}(\mathcal{D})).$$

It follows that

$$\begin{aligned}
\|\mathcal{J}(\mathcal{E}^{(\mathcal{U}_h(\hat{\xi},\hat{\Omega}))}))\|_p &\leq C\|\mathcal{U}_h(\hat{\xi},\hat{\Omega})\|_{L^p(J_T; W^{1+1/q+\varepsilon,q}(\mathcal{D}))} \\
&\leq CT^{1/18p}\|\mathcal{U}_h(\hat{\xi},\hat{\Omega})\|_{L^{\frac{18}{17}p}(J_T; W^{1+1/q+\varepsilon,q}(\mathcal{D}))} \\
&\leq CT^{1/18p}\|\mathcal{U}_h(\hat{\xi},\hat{\Omega})\|_{X^T_{p,q}} \\
&\leq CT^{1/18p}\|(\hat{\xi},\hat{\Omega})\|_{W^{1,p}(J_T)}. \tag{3.19}
\end{aligned}$$

Let $\mathcal{D}_R$ and $\mathcal{P}_h(\hat{\xi},\hat{\Omega})$ such that $\mathcal{P}_h(\hat{\xi},\hat{\Omega}) \in L^q_0(\mathcal{D}_R)$ and let again $\varepsilon > 0$ such that $c := \frac{1}{q} + \varepsilon < 1$. Then by interpolation, by the Poincar inequality and by (3.17),

$$\begin{aligned}
\|\mathcal{J}(\mathcal{P}_h(\hat{\xi},\hat{\Omega})\mathrm{Id})\|_{L^p(J_T)} &\leq C\|\mathcal{P}_h(\hat{\xi},\hat{\Omega})\|_{L^p(J_T; W^{1/q+\varepsilon,q}(\mathcal{D}_R))} \\
&\leq C\|\mathcal{P}_h(\hat{\xi},\hat{\Omega})\|^c_{L^p(J_T; L^q(\mathcal{D}_R))}\|\mathcal{P}_h(\hat{\xi},\hat{\Omega})\|^{1-c}_{L^p(J_T; W^{1,q}(\mathcal{D}_R))} \\
&\leq CT^{c\alpha/p}\|(\hat{\xi},\hat{\Omega})\|_{W^{1,p}(J_T)}\|\mathcal{P}_h(\hat{\xi},\hat{\Omega})\|^{1-c}_{Y^T_{p,q}} \\
&\leq CT^{c\alpha/p}\|(\hat{\xi},\hat{\Omega})\|_{W^{1,p}(J_T)}
\end{aligned}$$

for all $0 \leq \alpha < \frac{1}{2q}$. In conclusion,

$$\begin{aligned}
\|\mathcal{R}(\hat{\xi},\hat{\Omega})\|_{W^{1,p}(J_T)} &\leq C\|\mathcal{J}\mathbf{T}(\mathcal{U}_h(\hat{\xi},\hat{\Omega}) + \nabla v_{\hat{\xi},\hat{\Omega}}, \mathcal{P}_h(\hat{\xi},\hat{\Omega}))\|_p \\
&\leq C(T + T^{1/18p} + T^{c\alpha/p})\|(\hat{\xi},\hat{\Omega})\|_{W^{1,p}(J_T)}
\end{aligned}$$

and $\mathcal{R}(\hat{\xi}, \hat{\Omega})(0) = 0$ by definition, so that

$$L := \|\mathcal{R}\|_{\mathcal{L}(_0 W^{1,p}(J_T))} < 1 \tag{3.20}$$

for small T. Moreover,

$$\|f^*\|_{W^{1,p}(J_T)} \leq C \left\| \begin{pmatrix} f_1 \\ f_2 \end{pmatrix} - \mathcal{J}\mathbf{T}(u^*, p^*) \right\|_p$$
$$\leq C(\|f_0\|_{p,q} + \|(f_1, f_2)\|_p + \|u_0\|_{B^{2-2/p}_{q,p}(\mathcal{D})} + |(\xi_0, \Omega_0)|),$$

by (3.5). □

The lemma shows that for some $T > 0$, which depends on the geometry of the body but not on the initial data, the operator $\mathrm{Id} - \mathcal{R}$ is invertible. Thus, a unique solution to (3.16) exists on this time interval. Furthermore, we obtain the estimate

$$\|(\hat{\xi}, \hat{\Omega})\|_{W^{1,p}(J_T)} \leq (1 - L)C(\|f_0\|_{p,q} + \|(f_1, f_2)\|_p + \|u_0\|_{B^{2-2/p}_{q,p}(\mathcal{D})} + |(\xi_0, \Omega_0)|).$$

Plugging the solution $\hat{\xi}, \hat{\Omega}$ into (3.8), (3.13) and (3.14) yields solutions

$$\begin{aligned}
u &:= \hat{u} + \nabla v_{\hat{\xi}+\xi_0, \hat{\Omega}+\Omega_0} + u^* \in X^T_{p,q}, \\
p &:= \hat{p} - \partial_t v_{\hat{\xi}+\xi_0, \hat{\Omega}+\Omega_0} + p^* \in Y^T_{p,q}, \\
\xi &:= \hat{\xi} + \xi_0 \in W^{1,p}(J_T), \\
\Omega &:= \hat{\Omega} + \Omega_0 \in W^{1,p}(J_T)
\end{aligned}$$

of (3.1) and the estimate

$$\|u\|_{X^T_{p,q}} + \|p\|_{Y^T_{p,q}} + \|(\hat{\xi}, \hat{\Omega})\|_{W^{1,p}(J_T)}$$
$$\leq C(\|f_0\|_{p,q} + \|(f_1, f_2)\|_p + \|u_0\|_{B^{2-2/p}_{q,p}(\mathcal{D})} + |(\xi_0, \Omega_0)|).$$

The uniqueness of the solution for this linear problem immediately follows from the estimate. Since the length T of our time interval arises from condition (3.20) in the proof of Lemma 3.3, it is unaffected by the initial data and external forces and since $X^T_{p,q} \hookrightarrow C([0, T]; \{u \in B^{2-2/p}_{q,p}(\mathcal{D}) : \mathrm{div}\, u = 0\})$ and $W^{1,p}(J_T) \hookrightarrow C([0, T])$, we can take $u(T), \xi(T), \Omega(T)$ as initial values for solving the problem up to time $2T$. Iterating this procedure and gluing together the solutions on $(kT, (k+1)T)$ $k = 0, 1, 2, \ldots$ yields a solution on J_{T_0}. This proves Theorem 3.1.

Chapter 4

The Full Non-Linear Problem

In the first section of this chapter, we explain the structure of the fixed point argument and make preparations in notation. The technical work is in the second section, where we prove estimates on the transformed operators to show that the fixed point mapping is a contraction. In the last section, we give a precise formulation and a proof of the main result of the first four chapters regarding the existence of a unique local strong solution for problem (2.1).

4.1 Idea of the Fixed Point Argument

The maximal regularity estimate from the previous chapter is an important tool for tackling the non-linear problem (2.17). Similarly to the construction in [Tak03, Section 6], we introduce a fixed point problem in all four unknowns u, p, ξ and Ω.

First we deal with the gravitation term and enforce homogeneous initial conditions, so that the embedding constants can be fixed on varying time intervals. By Theorem 3.1, for every $T > 0$ there is a unique solution $u^*, p^*, \xi^*, \Omega^*$ to

$$
\begin{cases}
u_t^* - \Delta u^* + \nabla(p^* - \mathbf{g} \cdot y) = 0 & \text{in } J_T \times \mathcal{D}, \\
\operatorname{div} u^* = 0 & \text{in } J_T \times \mathcal{D}, \\
u^*(t,y) - \xi^*(t) - \Omega^*(t) \times y = 0, & t \in J_T, y \in \Gamma, \\
u^*(0) = v_0 & \text{in } \mathcal{D}, \\
\mathrm{m}(\xi^*)' + \int_\Gamma \mathbf{T}(u^*, p^*)N \, d\sigma = \mathrm{m}g & \text{in } J_T, \\
I(\Omega^*)' + \int_\Gamma y \times \mathbf{T}(u^*, p^*)N \, d\sigma = 0 & \text{in } J_T, \\
\xi^*(0) = \eta_0, \\
\Omega^*(0) = \omega_0,
\end{cases}
\tag{4.1}
$$

which satisfies

$$\|u^*\|_{X_{p,q}^T} + \|p^* - \mathbf{g} \cdot y\|_{Y_{p,q}^T} + \|(\xi^*, \Omega^*)\|_{W^{1,p}(J_T)}$$
$$\leq C(|\mathbf{mg}| + \|v_0\|_{B_{q,p}^{2-2/p}(D)} + |(\eta_0, \omega_0)|).$$

Let $\hat{u} = u - u^*$, $\hat{p} = p - (p^* - \mathbf{g} \cdot y)$, $\hat{\xi} = \xi - \xi^*$ and $\hat{\Omega} = \Omega - \Omega^*$. We rewrite problem (2.17) equivalently as

$$\begin{cases}
\hat{u}_t - \Delta \hat{u} + \nabla \hat{p} = F_0(\hat{u}, \hat{p}, \hat{\xi}, \hat{\Omega}) & \text{in } J_T \times \mathcal{D}, \\
\operatorname{div} \hat{u} = 0 & \text{in } J_T \times \mathcal{D}, \\
\hat{u}(t, y) - \hat{\xi}(t) - \hat{\Omega}(t) \times y = 0 & t \in J_T, y \in \Gamma, \\
\hat{u}(0) = 0, & \text{in } \mathcal{D}, \\
\mathrm{m}(\hat{\xi})' + \int_\Gamma \mathbf{T}(\hat{u}, \hat{p}) N \, d\sigma = F_1(\hat{u}, \hat{p}, \hat{\xi}, \hat{\Omega}) & \text{in } J_T, \\
I(\hat{\Omega})' + \int_\Gamma y \times \mathbf{T}(\hat{u}, \hat{p}) N \, d\sigma = F_2(\hat{u}, \hat{p}, \hat{\Omega}) & \text{in } J_T, \\
\hat{\xi}(0) = 0, & \\
\hat{\Omega}(0) = 0, &
\end{cases} \qquad (4.2)$$

where

$$F_0(\hat{u}, \hat{p}, \hat{\xi}, \hat{\Omega}) := \mathbf{G} - \mathbf{g} - \mathcal{N}(\hat{u} + u^*) + \mathcal{H}(\hat{u} + u^*, \hat{p} + p^*, \hat{\xi} + \xi^*, \hat{\Omega} + \Omega^*),$$
$$\mathcal{H}(u, p, \xi, \Omega) := -\mathcal{M}u + (\mathcal{L} - \Delta)u + (\mathcal{G} - \nabla)p,$$
$$F_1(\hat{u}, \hat{p}, \hat{\xi}, \hat{\Omega}) := \mathrm{m}(\mathbf{G}(\cdot, 0) - \mathbf{g}) - \mathrm{m}(\hat{\xi} + \xi^*) \times (\hat{\Omega} + \Omega^*)$$
$$+ \int_\Gamma (\mathbf{T} - \mathcal{T})(\hat{u}, \hat{p}) N \, d\sigma - \int_\Gamma \mathcal{T}(u^*, p^*) N \, d\sigma,$$
$$F_2(\hat{u}, \hat{p}, \hat{\Omega}) := -(\hat{\Omega} + \Omega^*) \times I(\hat{\Omega} + \Omega^*) + \int_\Gamma y \times (\mathbf{T} - \mathcal{T})(\hat{u}, \hat{p}) N \, d\sigma$$
$$- \int_\Gamma y \times \mathcal{T}(u^*, p^*) N \, d\sigma.$$

Note that the functions and operators $\mathcal{H}, \mathcal{N}, \mathbf{G}, \mathcal{T}$ themselves are not fixed, but they depend on the coordinate transform and therefore on the body velocity.

We can now formulate the fixed point argument. We look for a solution

$$\hat{u} \in X_{p,q,0}^T,$$
$$\hat{p} \in Y_{p,q}^T \text{ and}$$
$$\hat{\xi}, \hat{\Omega} \in {}_0W^{1,p}(J_T),$$

so we choose the length of the time interval T as a parameter and define the ball of radius R,

$$\mathcal{K}_R^T := \{(u, p, \xi, \Omega) \in X_{p,q,0}^T \times Y_{p,q}^T \times {}_0W^{1,p}(J_T; \mathbb{R}^6) :$$
$$\|u\|_{X_{p,q}^T} + \|p\|_{Y_{p,q}^T} + \|(\xi, \omega)\|_{W^{1,p}(J_T)} \leq R\},$$

as the underlying set. The exponents p, q should satisfy the assumption $\frac{3}{2q} + \frac{1}{p} \leq \frac{3}{2}$ motivated by the embedding properties of $X_{p,q}^T$ from Proposition 1.8. In the following, we denote the norm of functions f in the set $\mathcal{K}_R^T$ which is contained in the Banach space $X_{p,q}^T \times Y_{p,q}^T \times {}_0W^{1,p}(J_T)$ shortly by $\|f\|_{\mathcal{K}_R^T}$.

Let

$$\phi_R^T : \begin{pmatrix} \tilde{u} \\ \tilde{p} \\ \tilde{\xi} \\ \tilde{\Omega} \end{pmatrix} \mapsto \begin{pmatrix} F_0(\tilde{u}, \tilde{p}, \tilde{\xi}, \tilde{\Omega}) \\ F_1(\tilde{u}, \tilde{p}, \tilde{\xi}, \tilde{\Omega}) \\ F_2(\tilde{u}, \tilde{p}, \tilde{\Omega}) \end{pmatrix} \mapsto \begin{pmatrix} u \\ p \\ \xi \\ \Omega \end{pmatrix},$$

the function which maps $(\tilde{u}, \tilde{p}, \tilde{\xi}, \tilde{\Omega}) \in \mathcal{K}_R^T$ to the solution (u, p, ξ, Ω) of the linear problem (3.1) with right hand sides F_0, F_1, F_2 and initial values $\xi_0 = \Omega_0 = 0, u_0 = 0$. For sufficiently small $R, T > 0$, we show that the image of ϕ_R^T is contained in $\mathcal{K}_R^T$ and that the mapping is contractive. The fixed point of ϕ_R^T then satisfies (4.2).

In the next section, we use the estimates on the coordinate transforms X and Y from Chapter 2 to show that the coefficients and operators defining F_0, F_1 and F_2 yield contractions.

4.2 Estimates on the Transformed Operators

We fix $T_0, R_0 > 0$. In the following, $C > 0$ denotes a generic constant which does not depend on the parameters T, R for $0 < T \leq T_0$, $0 < R \leq R_0$. For example, factors of the form e^{RT} or $T^{1/p}$ will be replaced by their upper bounds $e^{R_0 T_0}$ and $T_0^{1/p}$. So in order to keep the estimates perspicuous, they are not stated sharply. We set

$$C_0 := \|u^*\|_{X_{p,q}^{T_0}} + \|p^* - \mathbf{g} \cdot y\|_{Y_{p,q}^{T_0}} + \|(\xi^*, \Omega^*)\|_{W^{1,p}(I_{T_0})}.$$

From now on, we always assume $(\tilde{u}, \tilde{p}, \tilde{\xi}, \tilde{\Omega}), (\tilde{u}_1, \tilde{p}_1, \tilde{\xi}_1, \tilde{\Omega}_1), (\tilde{u}_2, \tilde{p}_2, \tilde{\xi}_2, \tilde{\Omega}_2) \in \mathcal{K}_R^T$ and that $u^*, p^*, \xi^*, \Omega^*$ are given by (4.1). We abuse our notation by setting $u = \tilde{u} + u^*, p = \tilde{p} + p^*, \xi = \tilde{\xi} + \xi^*, \Omega = \tilde{\Omega} + \Omega^*$ and $u_1 = \tilde{u}_1 + u^*, p_1 = \tilde{p}_1 + p^*$ etc. throughout this section. Note that we often use the estimate

$$\begin{aligned} &\|(u, p, \xi, \Omega)\|_{X_{p,q}^T \times Y_{p,q}^T \times W^{1,p}(J_T)} \\ \leq\ & \|(\tilde{u}, \tilde{p}, \tilde{\xi}, \tilde{\Omega})\|_{\mathcal{K}_R^T} + \|(u^*, p^*, \xi^*, \Omega^*)\|_{X_{p,q}^T \times Y_{p,q}^T \times W^{1,p}(J_T)} \\ \leq\ & R + C_0. \end{aligned}$$

Given ξ, ξ_1, ξ_2 and $\Omega, \Omega_1, \Omega_2$ from $\tilde{\xi}, \tilde{\xi}_1, \tilde{\xi}_2, \xi^*$ and $\tilde{\Omega}, \tilde{\Omega}_1, \tilde{\Omega}_2, \Omega^*$, we can calculate the corresponding coordinate transforms X, X_1, X_2 and Y, Y_1, Y_2 in the way described in Section 2.3, steps (1) - (5). They determine the covariant and the contravariant metric tensor and the Christoffel symbol, cf. (2.14), (2.15) and (2.16), we used in the formulation of the transformed problem (2.17). These coefficients $g^{ij}, g_{ij}, \partial_l g_{ij}, \partial_l g^{ij}, \Gamma^i_{jk}$ and $\partial_l \Gamma^i_{jk}$ satisfy the uniform estimates in the following lemma.

Lemma 4.1. *Let $T > 0$. The coefficients given by the coordinate transforms X, X_1, X_2 and Y, Y_1, Y_2 satisfy*

$$\|\partial^\alpha g^{ij}\|_{\infty,\infty} + \|\partial^\alpha g_{ij}\|_{\infty,\infty} + \|\partial^\alpha \Gamma^i_{jk}\|_{\infty,\infty} \leq C$$

and

$$\|\partial^\alpha((g_1)^{ij} - (g_2)^{ij})\|_{\infty,\infty} + \|\partial^\alpha((g_1)_{ij} - (g_2)_{ij})\|_{\infty,\infty} + \|\partial^\alpha((\Gamma_1)^i_{jk} - (\Gamma_2)^i_{jk})\|_{\infty,\infty}$$
$$\leq CT \left\| (\tilde\xi_1 - \tilde\xi_2, \tilde\Omega_1 - \tilde\Omega_2) \right\|_\infty$$

for all multi-indices $0 \leq |\alpha| \leq 1$.

Proof. The estimates are a direct consequence of Proposition 2.1. Since $g_{ij}(t,y) = \Sigma^3_{l=1}(\partial_l X_i)(\partial_l X_j)(t,y)$,

$$\|g_{ij}\|_{\infty,\infty} \leq C \sup_l \|\partial_l X\|^2_{\infty,\infty} \leq C$$

and the first order partial derivatives satisfy

$$\|\partial_k g_{ij}\|_{\infty,\infty} \leq C \sup_l \|\partial_k \partial_l X\|_{\infty,\infty} \|\partial_l X\|_{\infty,\infty} \leq C.$$

In Lemma 2.1, the constant C depends on $\|(\xi,\Omega)\|_\infty$, but here we can additionally use

$$\|(\xi,\Omega)\|_\infty \leq \|(\tilde\xi,\tilde\Omega)\|_\infty + \|(\xi^*,\Omega^*)\|_{L^\infty(J_{T_0})}$$
$$\leq C_1\|(\tilde\xi,\tilde\Omega)\|_{W^{1,p}(J_T)} + C_2 \|(\xi^*,\Omega^*)\|_{W^{1,p}(J_{T_0})}$$
$$\leq C(R + C_0),$$

where C_1 is chosen independently of T because $\tilde\xi, \tilde\Omega \in {}_0W^{1,p}(J_T)$, so that this dependence is eliminated. Furthermore, the differences of two coefficients given by two different transforms satisfy

$$\|(g_1)_{ij} - (g_2)_{ij}\|_{\infty,\infty} \leq C \sup_l (\|\partial_l X_1\|_{\infty,\infty} + \|\partial_l X_2\|_{\infty,\infty}) \|\partial_l (X_1 - X_2)\|_{\infty,\infty}$$
$$\leq CT\|(\tilde\xi_1 - \tilde\xi_2, \tilde\Omega_1 - \tilde\Omega_2)\|_\infty.$$

The remaining estimates follow analogously.   □

With these estimates on the coefficients, we can control the transformed differential operators appearing in F_0.

Lemma 4.2. *Let $T_0 \geq T > 0$ and $\frac{3}{2q} + \frac{1}{p} \leq \frac{3}{2}$. Let furthermore $s = 3p, s' = 3p/2, r = 3q, r' = 3q/2$ and*

$$C_*(T) := \|\nabla u^*\|_{s',r'} + \|u^*\|_{s,r}.$$

Then

$$
\begin{aligned}
\|\mathbf{G} - \mathbf{g}\|_{p,q} &\leq CT \|(\xi, \Omega)\|_{W^{1,p}(J_T)}, \\
\|\mathcal{M}u\|_{p,q} &\leq C(T^{1/2} + T^{1/p}) \|u\|_{X_{p,q}^T}, \\
\|(\mathcal{L} - \Delta)u\|_{p,q} &\leq C(T + T^{1/2} + T^{1/p}) \|u\|_{X_{p,q}^T}, \\
\|(\mathcal{G} - \nabla)p\|_{p,q} &\leq CT \|p\|_{Y_{p,q}^T}, \\
\|\mathcal{N}(u)\|_{p,q} &\leq C[(R + C_*(T))^2 + T^{1/2}].
\end{aligned}
$$

Moreover, we obtain

$$\|\mathcal{J}((\mathbf{T} - \mathcal{T})(\tilde{u}, \tilde{p}))\|_p \leq CT(\|\tilde{u}\|_{X_{p,q}^T} + \|\tilde{p}\|_{Y_{p,q}^T}).$$

Proof. These estimates follow from Lemma 4.1 and the embedding properties of $X_{p,q}^T$.

If $p \geq 2$, let $k' = \infty$ and fix $\frac{1}{k'} = \frac{1}{p} - \frac{1}{2}$ otherwise. Proposition 1.8 yields the embeddings

$$X_{p,q}^T \hookrightarrow L^s(J_T; L^r(\mathcal{D})), \quad X_{p,q}^T \hookrightarrow L^{s'}(J_T; W^{1,r'}(\mathcal{D})) \quad \text{and}$$

$$X_{p,q}^T \hookrightarrow L^{k'}(J_T; W^{1,q}(\mathcal{D})),$$

where the embedding constants do not depend on T provided $\tilde{u} \in X_{p,q,0}^T$ and $u^* \in X_{p,q}^{T_0}$.

There are two operators of highest order, $\mathcal{L}$ and $\mathcal{G}$, which differ from Δ and ∇ smoothly in time. The identity transform $X(t,y) = y$ for all $t > 0, y \in \mathbb{R}^3$ corresponds to the body velocities $\xi = \Omega = 0$, so

$$
\begin{aligned}
\|g^{jk} - \delta_{jk}\|_{\infty,\infty} &\leq CT \|(\xi, \Omega)\|_{W^{1,p}(J_T)} \quad \text{and} \\
\|\partial_j Y_k - \delta_{jk}\|_{\infty,\infty} &\leq CT \|(\xi, \Omega)\|_{W^{1,p}(J_T)}
\end{aligned}
$$

by Lemma 4.1. It follows that

$$\|\mathbf{G} - \mathbf{g}\|_{p,q} \leq C \sup_{j,k} \|\partial_j Y_k - \delta_{jk}\|_{\infty,\infty} |\mathbf{g}| \leq CT \|(\xi, \Omega)\|_{W^{1,p}(J_T)}$$

and that

$$
\begin{aligned}
&\|(\mathcal{L} - \Delta)u\|_{p,q} \\
\leq\ & C \sup_{i,j,k,l} \big[\|g^{jk} - \delta_{jk}\|_{\infty,\infty} \|\Delta u\|_{p,q} \\
& + \big(\|\partial_j g^{jk}\|_{\infty,\infty} + \|g^{ki}\|_{\infty,\infty} \|\Gamma^i_{jk}\|_{\infty,\infty} \big) \|\nabla u\|_{p,q} \\
& + \big(\|\partial_k g^{kl}\|_{\infty,\infty} \|\Gamma^i_{kl}\|_{\infty,\infty} + \|g^{kl}\|_{\infty,\infty} \|\partial_k \Gamma^i_{kl}\|_{\infty,\infty} \\
& + \|g^{kl}\|_{\infty,\infty} \|\Gamma^m_{jl}\|_{\infty,\infty} \|\Gamma^i_{km}\|_{\infty,\infty} \big) \|u\|_{p,q} \big] \\
\leq\ & CT(R + C_0) \|u\|_{X^T_{p,q}} + CT^{1/p - 1/k'} \|u\|_{L^{k'}(J_T; W^{1,q}(\mathcal{D}))} \\
& + CT^{1/p} \|u\|_{\infty,q} \\
\leq\ & C(T + T^{1/2} + T^{1/p})(R + C_0) \|u\|_{X^T_{p,q}}
\end{aligned}
$$

as well as

$$
\|(\mathcal{G} - \nabla)p\|_{p,q} \ \leq\ C \sup_{j,k} \|g^{jk} - \delta_{jk}\|_{\infty,\infty} \|\nabla p\|_{p,q} \leq CT \|p\|_{Y^T_{p,q}} \, .
$$

The additional terms arising from the material derivative satisfy

$$
\begin{aligned}
\|\mathcal{M}u\|_{p,q} \ \leq\ & C \|b^{(Y)}\|_{\infty,\infty} \|\nabla u\|_{p,q} + \sup_{i,j,k} \big(\|\Gamma^i_{jk}\|_{\infty,\infty} \|b^{(Y)}\|_{\infty,\infty} \\
& + \|\partial_k Y_i(\cdot, X)\|_{\infty,\infty} \|J_b\|_{\infty,\infty} \big) \|u\|_{p,q} \\
\leq\ & CT^{1/2} \|u\|_{L^{k'}(J_T; W^{1,q}(\mathcal{D}))} + CT^{1/p} \|u\|_{\infty,q} \\
\leq\ & C(T^{1/2} + T^{1/p}) \|u\|_{X^T_{p,q}}
\end{aligned}
$$

by Proposition 2.1 and Lemmas 2.5, 2.3 and 2.2. For the transformed convection term, we use the fact that $X^T_{p,q} \hookrightarrow L^l(J_T; L^{2q}(\mathcal{D}))$ for $\frac{1}{l} := \frac{1}{p} + \frac{3}{4q} - 1$. Thus,

$$
\begin{aligned}
\|\mathcal{N}(u)\|_{p,q} \ \leq\ & \|(u \cdot \nabla)u)\|_{p,q} + \sup_{i,j,k} \|\Gamma^i_{jk}\|_{\infty,\infty} \|u\|^2_{2p,2q} \\
\leq\ & \|u\|_{r,s} \|\nabla u\|_{r',s'} + CT^{2/2p - 2/l} \|u\|^2_{L^l(J_T; L^{2q}(\mathcal{D}))} \\
\leq\ & C[(R + C_*(T))^2 + T^{1/2}].
\end{aligned}
$$

Concerning the non-linear term in the rigid body equations, we can set $M_2 = 0$ in (2.19) to see that the estimate

$$
\|Q - \mathrm{Id}_{\mathbb{R}^3}\|_\infty \leq CT \|\Omega\|_\infty
$$

holds true for the matrix Q associated to $\tilde{\Omega} + \Omega^*$ via $Q' = MQ$, $Q(0) = \mathrm{Id}_{\mathbb{R}^3}$ as in (2.13), where M is the matrix-valued function satisfying $(\tilde{\Omega} + \Omega^*)(t) \times y =$

$M(t)y$ for all $t \in J_T, y \in \mathbb{R}^3$. This means that by (3.18) and (2.19),

$$\begin{aligned}
\|\mathcal{J}(\tilde{p}\mathrm{Id} - Q^T\tilde{p}Q)\|_p &\leq C(\|(\mathrm{Id}_{\mathbb{R}^3} - Q^T)\tilde{p}\|_{Y_{p,q}^T} + \|Q^T\tilde{p}(\mathrm{Id}_{\mathbb{R}^3} - Q)\|_{Y_{p,q}^T}) \\
&\leq CT\|\tilde{p}\|_{Y_{p,q}^T}.
\end{aligned} \tag{4.3}$$

Similarly, we show

$$\begin{aligned}
\|\mathcal{J}(\mathcal{E}^{(\tilde{u})} - Q^T\mathcal{E}^{(Q\tilde{u})}Q)\|_p &\leq \|(\mathrm{Id}_{\mathbb{R}^3} - Q^T)\mathcal{E}^{(\tilde{u})})\|_{L^p(J_T;W^{1,q}(\mathcal{D}))} \\
&\quad + \|Q^T\mathcal{E}^{(\tilde{u}-Q\tilde{u})}\|_{L^p(J_T;W^{1,q}(\mathcal{D}))} \\
&\quad + \|Q^T\mathcal{E}^{(\tilde{u})}(\mathrm{Id}_{\mathbb{R}^3} - Q)\|_{L^p(J_T;W^{1,q}(\mathcal{D}))} \\
&\leq CT\|\tilde{u}\|_{X_{p,q}^T},
\end{aligned}$$

where we use $\mathcal{E}^{(Qu)} = \frac{1}{2}(Q\nabla u + (\nabla u)^T Q^T)$ and put $\varepsilon = 1 - 1/q$ in (3.18). $\square$

We now have the main estimates at hand to prove the following lemma.

Lemma 4.3. *Given $T, R > 0$ sufficiently small, the function ϕ_R^T maps $\mathcal{K}_R^T$ into itself.*

Proof. Clearly by Lemma 4.2,

$$\begin{aligned}
\|F_1(\tilde{u}, \tilde{p}, \tilde{\xi}, \tilde{\Omega})\|_p &\leq \|\mathbf{G}(\cdot, 0) - \mathbf{g}\|_p + \|\xi \times \Omega\|_p + \|\mathcal{J}((\mathbf{T} - \mathcal{T})(\hat{u}, \hat{p}))\|_p \\
&\leq \|Q^T - \mathrm{Id}_{\mathbb{R}^3}\|_p |\mathbf{g}| + \|\xi\|_\infty \|\Omega\|_p + CTR \\
&\leq CT + T^{1/p}(R + C_0) + CTR \\
&\leq CT^{1/p}
\end{aligned}$$

and

$$\|F_2(\tilde{u}, \tilde{p}, \tilde{\Omega})\|_p \leq \|\Omega \times I\Omega\|_p + \|\mathcal{J}((\mathbf{T} - \mathcal{T})(\tilde{u}, \tilde{p}))\|_p \leq CT^{1/p}.$$

Furthermore, by Lemma 4.2 we also get

$$\begin{aligned}
\|F_0(\tilde{u}, \tilde{p}, \tilde{\xi}, \tilde{\Omega})\|_{p,q} &\leq \|\mathbf{G} - \mathbf{g}\|_{p,q} + \|\mathcal{H}(u, p)\|_{p,q} + \|\mathcal{N}(u)\|_{p,q} \\
&\leq C[T^{1/2} + T^{1/p} + (R + C_*(T))^2].
\end{aligned}$$

Note here that $C_*(T) \to 0$ as $T \to 0$ by the definition in Lemma 4.2. Thus by Theorem 3.1,

$$\begin{aligned}
\|\phi_R^T(\tilde{u}, \tilde{p}, \tilde{\xi}, \tilde{\Omega})\|_{\mathcal{K}_R^T} &\leq C(\|F_0(\tilde{u}, \tilde{p}, \tilde{\xi}, \tilde{\Omega})\|_{p,q} \\
&\quad + \|F_1(\tilde{u}, \tilde{p}, \tilde{\xi}, \tilde{\Omega})\|_p + \|F_2(\tilde{u}, \tilde{p}, \tilde{\Omega})\|_p) \\
&\leq C(T^{1/2} + T^{1/p} + (R + C_*(T))^2) \\
&\leq R,
\end{aligned}$$

if we choose T and R sufficiently small. $\square$

In a similar way, we show

Lemma 4.4. *The map ϕ_R^T is contractive.*

Proof. Again, the proof is mainly a matter of writing out the estimates on F_0. We show that

$$\|F_0(\tilde{u}_1, \tilde{p}_1, \tilde{\xi}_1, \tilde{\Omega}_1) - F_0(\tilde{u}_2, \tilde{p}_2, \tilde{\xi}_2, \tilde{\Omega}_2)\|_{p,q}$$
$$\leq L_{R,T}\|(\tilde{u}_1 - \tilde{u}_2, \tilde{p}_1 - \tilde{p}_2, \tilde{\xi}_1 - \tilde{\xi}_2, \tilde{\Omega}_1 - \tilde{\Omega}_2)\|_{\mathcal{K}_R^T}, \tag{4.4}$$

where $L_{R,T}$ can be made arbitrarily small for $T, R \to 0$. First, we rewrite the term in the following form.

$$\begin{aligned}
F_0(\tilde{u}_1, \tilde{p}_1, \tilde{\xi}_1, \tilde{\Omega}_1) - F_0(\tilde{u}_2, \tilde{p}_2, \tilde{\xi}_2, \tilde{\Omega}_2) &= \mathcal{H}_1(\tilde{u}_1 - \tilde{u}_2, \tilde{p}_1 - \tilde{p}_2, \tilde{\xi}_1 - \tilde{\xi}_2, \tilde{\Omega}_1 - \tilde{\Omega}_2) \\
&\quad + \mathbf{G}_1 - \mathbf{G}_2 + (\mathcal{M}_1 - \mathcal{M}_2)u_2 \\
&\quad + (\mathcal{L}_1 - \mathcal{L}_2)u_2 + (\mathcal{G}_1 - \mathcal{G}_2)p_2 \\
&\quad + \mathcal{N}_1(u_1) - \mathcal{N}_2(u_2).
\end{aligned}$$

The estimates for $\mathcal{H}_1$ are already known from Lemma 4.2. Concerning the remaining terms, we show that by Proposition 2.1 and by Lipschitz continuity of the Jacobians J_{Y_i} of Y_i, $i \in \{1, 2\}$,

$$\begin{aligned}
\|\mathbf{G}_1 - \mathbf{G}_2\|_{p,q} &\leq C\|J_{Y_1}(\cdot, X_1) - J_{Y_2}(\cdot, X_2)\|_{p,q} \\
&\leq CT\|(\tilde{\xi}_1 - \tilde{\xi}_2, \tilde{\Omega}_1 - \tilde{\Omega}_2)\|_{W^{1,p}(J_T)}.
\end{aligned}$$

By Lemma 4.1,

$$\begin{aligned}
&\|(\mathcal{M}_1 - \mathcal{M}_2)u_2\|_{p,q} \\
&\leq C\|b_1^{(Y)} - b_2^{(Y)}\|_{\infty,\infty}\|\nabla u_2\|_{p,q} \\
&\quad + \sup_{i,j,k}\big(\|(\Gamma_1)_{jk}^i - (\Gamma_2)_{jk}^i\|_{\infty,\infty}\|b_2^{(Y)}\|_{\infty,\infty} + \|(\Gamma_2)_{jk}^i\|_{\infty,\infty}\|b_1^{(Y)} - b_2^{(Y)}\|_{\infty,\infty} \\
&\quad + \|J_{Y_1}(\cdot, X_1) - J_{Y_2}(\cdot, X_2)\|_{\infty,\infty}\|J_{b_1}\|_{\infty,\infty} \\
&\quad + \|J_{Y_2}(\cdot, X_2)\|_{\infty,\infty}\|J_{b_1} - J_{b_2}\|_{\infty,\infty}\big)\|u_2\|_{p,q} \\
&\leq CT(R + C_0)\|(\tilde{\xi}_1 - \tilde{\xi}_2, \tilde{\Omega}_1 - \tilde{\Omega}_2)\|_{W^{1,p}(J_T)},
\end{aligned}$$

and similarly,

$$\|(\mathcal{L}_1 - \mathcal{L}_2)u_2\|_{p,q} \leq CT\|(\tilde{\xi}_1 - \tilde{\xi}_2, \tilde{\Omega}_1 - \tilde{\Omega}_2)\|_{W^{1,p}(J_T)} \tag{4.5}$$

and

$$\|(\mathcal{G}_1 - \mathcal{G}_2)p_2\|_{p,q} \leq CT\|(\tilde{\xi}_1 - \tilde{\xi}_2, \tilde{\Omega}_1 - \tilde{\Omega}_2)\|_{W^{1,p}(J_T)}.$$

Moreover,

$$\|\mathcal{N}_1(u_1) - \mathcal{N}_2(u_2)\|_{p,q}$$
$$\leq \|(u_1 \cdot \nabla)u_1 - (u_2 \cdot \nabla)u_2\|_{p,q}$$
$$+ \sup_i \|\sum_{j,k}\left((\Gamma_1)^i_{jk}(u_1)_j(u_1)_k - (\Gamma_2)^i_{jk}(u_2)_j(u_2)_k\right)\|_{p,q}$$
$$\leq C \sup_{i,j,k}(\|u_1\|_{s,r}\|\nabla(\tilde{u}_1 - \tilde{u}_2)\|_{s',r'} + \|\nabla u_2\|_{s',r'}\|\tilde{u}_1 - \tilde{u}_2\|_{s,r}$$
$$+ \|(\Gamma_1)^i_{jk}\|_{\infty,\infty}(\|u_1\|_{2p,2q} + \|u_2\|_{2p,2q})\|\tilde{u}_1 - \tilde{u}_2\|_{2p,2q}$$
$$+ \|(\Gamma_1)^i_{jk} - (\Gamma_2)^i_{jk}\|_{\infty,\infty}\|u_2\|^2_{2p,2q}$$
$$\leq C(R + C_*(T))(\|\tilde{u}_1 - \tilde{u}_2\|_{X^T_{p,q}} + \|(\tilde{\xi}_1 - \tilde{\xi}_2, \tilde{\Omega}_1 - \tilde{\Omega}_2)\|_{W^{1,p}(J_T)}),$$

where s, s', r and r' are defined as in Lemma 4.3. In conclusion, (4.4) holds true if we set $L_{R,T} := C(R + C_*(T) + T + T^{1/2} + T^{1/p})$ and use that $C_*(T) \to 0$ as $T \to 0$. Concerning the functions F_1, F_2, we first show that

$$\|\mathcal{J}_1 - \mathcal{J}_2\|_p := \|\mathcal{J}((\mathbf{T} - \mathcal{T}_1)(\tilde{u}_1, \tilde{p}_1) - (\mathbf{T} - \mathcal{T}_2)(\tilde{u}_2, \tilde{p}_2))\|_p$$
$$\leq \|\mathcal{J}((\mathbf{T} - \mathcal{T}_1)(\tilde{u}_1 - \tilde{u}_2, \tilde{p}_1 - \tilde{p}_2))\|_p + \|\mathcal{J}((\mathcal{T}_1 - \mathcal{T}_2)(\tilde{u}_2, \tilde{p}_2))\|_p$$
$$\leq CT(\|\tilde{u}_1 - \tilde{u}_2\|_{X^T_{p,q}} + \|\tilde{p}_1 - \tilde{p}_2\|_{Y^T_{p,q}} + \|\tilde{\Omega}_1 - \tilde{\Omega}_2\|_{W^{1,p}(J_T)}).$$

The first term in the second line satisfies

$$\|\mathcal{J}((\mathbf{T} - \mathcal{T}_1)(\tilde{u}_1 - \tilde{u}_2, \tilde{p}_1 - \tilde{p}_2))\|_p \leq CT(\|\tilde{u}_1 - \tilde{u}_2\|_{X^T_{p,q}} + \|\tilde{p}_1 - \tilde{p}_2\|_{Y^T_{p,q}})$$

by Lemma 4.2. The estimate for the second term can be shown in a similar way, again using the estimates on $\mathcal{J}$ proved in Section 3.2 and (2.19), where $Q_i \in W^{1,p}(J_T; \mathbb{R}^{3\times3})$ are associated to the body velocities $\tilde{\xi}_i + \xi^*, \tilde{\Omega}_i + \Omega^*$ for $i \in \{1, 2\}$ as in Section 2.3,

$$\|\mathcal{J}((\mathcal{T}_1 - \mathcal{T}_2)(\tilde{u}_2, \tilde{p}_2))\|_p$$
$$\leq CT^{\frac{1}{18p}}\left[\|Q_1^T - Q_2^I\|_\infty\|Q_1\|_\infty\|\tilde{u}_2\|_{X^T_{p,q}}\|Q_1\|_\infty\right.$$
$$+ \|Q_2^T\|_\infty\|(Q_1^T - Q_2^T)\|_\infty\|\tilde{u}_2\|_{X^T_{p,q}}\|Q_1\|_\infty$$
$$+ \|Q_2^T\|_\infty\|Q_2\|_\infty\|\tilde{u}_2\|_{X^T_{p,q}}\|Q_1 - Q_2\|_\infty\right]$$
$$+ C\left[\|Q_1^T - Q_2^T\|_\infty\|\tilde{p}_2\|_{Y^T_{p,q}}\|Q_1\|_\infty + \|Q_2^T\|_\infty\|\tilde{p}_2\|_{Y^T_{p,q}}\|Q_1 - Q_2\|_\infty\right]$$
$$\leq CT\|\tilde{\Omega}_1 - \tilde{\Omega}_2\|_{W^{1,p}(J_T)}. \tag{4.6}$$

In conclusion,

$$\|F_1(\tilde{u}_1, \tilde{p}_1, \tilde{\xi}_1, \tilde{\Omega}_1) - F_1(\tilde{u}_2, \tilde{p}_2, \tilde{\xi}_2, \tilde{\Omega}_2)\|_p$$
$$\leq \|\mathbf{G}_1(\cdot, 0) - \mathbf{G}_2(\cdot, 0)\|_p + \|(\tilde{\xi}_1 + \xi^*) \times (\tilde{\Omega}_1 - \tilde{\Omega}_2)\|_p$$
$$+ \|(\tilde{\xi}_1 - \tilde{\xi}_2) \times (\tilde{\Omega}_2 + \Omega^*)\|_p + \|\mathcal{J}_1 - \mathcal{J}_2\|_p$$

$$\leq\ C\,\|Q_1 - Q_2\|_p + (R + C_0)\|(\tilde{\xi}_1 - \tilde{\xi}_2, \tilde{\Omega}_1 - \tilde{\Omega}_2)\|_p$$
$$+ CT(\|\tilde{u}_1 - \tilde{u}_2\|_{X^T_{p,q}} + \|\tilde{p}_1 - \tilde{p}_2\|_{Y^T_{p,q}} + \|\tilde{\Omega}_1 - \tilde{\Omega}_2\|_{W^{1,p}(J_T)})$$
$$\leq\ CT\|(\tilde{u}_1 - \tilde{u}_2, \tilde{p}_1 - \tilde{p}_2, \tilde{\xi}_1 - \tilde{\xi}_2, \tilde{\Omega}_1 - \tilde{\Omega}_2)\|_{\mathcal{K}^T_R}$$

and

$$\|F_2(\tilde{u}_1, \tilde{p}_1, \tilde{\Omega}_1) - F_2(\tilde{u}_2, \tilde{p}_2, \tilde{\Omega}_2)\|_p$$
$$\leq\ \|(\tilde{\Omega}_1 + \Omega^*) \times I(\tilde{\Omega}_1 - \tilde{\Omega}_2)\|_p + \|(\tilde{\Omega}_1 - \tilde{\Omega}_2) \times I(\tilde{\Omega}_2 + \Omega^*)\|_p + \|\mathcal{J}_1 - \mathcal{J}_2\|_p$$
$$\leq\ CT(\|\tilde{u}_1 - \tilde{u}_2\|_{X^T_{p,q}} + \|\tilde{p}_1 - \tilde{p}_2\|_{Y^T_{p,q}} + \|\tilde{\Omega}_1 - \tilde{\Omega}_2\|_{W^{1,p}(J_T)}).$$

By Theorem 3.1, it follows from these estimates that ϕ^T_R is strongly contractive for sufficiently small T, R, due to

$$\|\phi^T_R(\tilde{u}_1, \tilde{p}_1, \tilde{\xi}_1, \tilde{\Omega}_1) - \phi^T_R(\tilde{u}_2, \tilde{p}_2, \tilde{\xi}_2, \tilde{\Omega}_2)\|_{\mathcal{K}^T_R}$$
$$\leq\ C\big(\|F_0(\tilde{u}_1, \tilde{p}_1, \tilde{\xi}_1, \tilde{\Omega}_1) - F_0(\tilde{u}_2, \tilde{p}_2, \tilde{\xi}_2, \tilde{\Omega}_2)\|_{p,q}$$
$$+ \|F_1(\tilde{u}_1, \tilde{p}_1, \tilde{\xi}_1, \tilde{\Omega}_1) - F_1(\tilde{u}_2, \tilde{p}_2, \tilde{\xi}_2, \tilde{\Omega}_2)\|_p$$
$$+ \|F_2(\tilde{u}_1, \tilde{p}_1, \tilde{\Omega}_1) - F_2(\tilde{u}_2, \tilde{p}_2, \tilde{\Omega}_2)\|_p\big)$$
$$\leq\ CL_{R,T}\|(\tilde{u}_1 - \tilde{u}_2, \tilde{p}_1 - \tilde{p}_2, \tilde{\xi}_1 - \tilde{\xi}_2, \tilde{\Omega}_1 - \tilde{\Omega}_2)\|_{\mathcal{K}^T_R}.$$

$\square$

4.3   Main Result

The contraction mapping theorem yields a unique fixed point of ϕ^T_R which is a strong solution $(\hat{u}, \hat{p}, \hat{\xi}, \hat{\Omega})$ of problem (4.2). The solution (u, p, ξ, Ω) of the transformed problem (2.17) can be obtained by adding the reference solution,

$$u = \hat{u} + u^*, \quad p = \hat{p} + p^* - \mathbf{g} \cdot y \quad \xi = \hat{\xi} + \xi^*, \quad \Omega = \hat{\Omega} + \Omega^*.$$

It was shown in Chapter 2 that this gives us a solution (v, q, η, ω) to the original free-fall problem (2.1) if we set

$$\begin{aligned} v(t,x) &= J_X(t, Y(t,x))u(t, Y(t,x)), \\ q(t,x) &= p(t, Y(t,x)), \\ \eta(t) &= Q(t)\xi(t), \\ \omega(t) &= Q(t)\Omega(t). \end{aligned}$$

So far, we only stated that the transform X preserves regularity. To make this more precise, we define the spaces of maximal regularity on the moving domain $\mathcal{D}(t)$ in the following way.

Definition 4.5. For $T > 0$ let

$$L^p(J_T; L^q(\mathcal{D}(\cdot))) := \{f \in L^1_{\mathrm{loc}}(J_T \times \mathcal{D}(\cdot)) : \|f\|_{L^p(J_T; L^q(\mathcal{D}(\cdot)))} < \infty\},$$

where

$$\|f\|_{L^p(J_T; L^q(\mathcal{D}(\cdot)))} := \left(\int_0^T \|f(t)\|^p_{L^q(\mathcal{D}(t))} \, \mathrm{d}t \right)^{1/p}.$$

Analogously, we define the spaces $L^p(J_T; W^{2,q}(\mathcal{D}(\cdot)))$, $W^{1,p}(J_T; L^q(\mathcal{D}(\cdot)))$, $L^p(J_T; \widehat{W}^{1,q}(\mathcal{D}(\cdot)))$ and $C^1(J_T; C^\infty(\mathcal{D}(\cdot)))$.

It is easy to see that v and q are contained in these spaces. More precisely, the following theorem is the main result of this chapter.

Theorem 4.6. *Let $1 < p, q < \infty$ such that $\frac{3}{2q} + \frac{1}{p} \leq \frac{3}{2}$ and $\frac{1}{2q} + \frac{1}{p} \neq 1$ and let $\eta_0, \omega_0 \in \mathbb{R}^3$ and $v_0 \in B_{q,p}^{2-2/p}(\mathcal{D})$ such that it satisfies the compatibility conditions (3.2) and (3.3). Then there exists a maximal interval $[0, T_0)$, $T_0 > 0$ such that problem (2.1) admits a unique strong solution*

$$\begin{aligned}
v &\in L^p(J_{T_0}; W^{2,q}(\mathcal{D}(\cdot))) \cap W^{1,p}(J_{T_0}; L^q(\mathcal{D}(\cdot))), \\
q &= q_0 + \mathbf{g} \cdot Y, \quad q_0 \in L^p(J_{T_0}; \widehat{W}^{1,q}(\mathcal{D}(\cdot))), Y \in C^1(J_{T_0}; C^\infty(\mathcal{D}(\cdot))), \\
\eta &\in W^{1,p}(J_{T_0}), \\
\omega &\in W^{1,p}(J_{T_0}).
\end{aligned}$$

The maximal interval of existence of solutions is characterized as follows. Either $T_0 = +\infty$ or one of the functions

$$t \mapsto \|v(t)\|_{B_{q,p}^{2-2/p}(\mathcal{D}(t))}, \qquad t \mapsto |\eta(t)|, \qquad t \mapsto |\omega(t)|$$

is unbounded on $[0, T_0)$, because if T_0 were finite and the above maps bounded, we could use $v(T_0), \eta(T_0), \omega(T_0)$ as new initial values for our problem to obtain a solution $v_\varepsilon, q_\varepsilon, \eta_\varepsilon, \omega_\varepsilon$ on some time interval $(T_0, T_0 + \varepsilon)$ as in the proof of Theorem 3.1. Extending v, q, η, ω by $v_\varepsilon, q_\varepsilon, \eta_\varepsilon, \omega_\varepsilon$ yields a contradiction.

It remains to argue that the solution v, q, η, ω is unique. This is basically a consequence of the uniqueness of the fixed point $\hat{u}, \hat{p}, \hat{\xi}, \hat{\Omega}$ which solves (4.2). Given η_0, ω_0, v_0 and $\mathbf{g} \in \mathbb{R}^3$, we assume that both $s_1 := (v_1, q_1, \eta_1, \omega_1)$ and $s_2 := (v_2, q_2, \eta_2, \omega_2)$ solve (2.1) on the time interval J_{T_0}. If we choose some cut-off function χ, cf. (2.6), we can transform s_1 and s_2 to the functions $t_1 := (u_1, p_1, \xi_1, \Omega_1)$ and $t_2 := (u_2, p_2, \xi_2, \Omega_2)$ as defined in (2.4) and (2.12). The reference solution $t_* := (u^*, p^* - \mathbf{g} \cdot y, \xi^*, \Omega^*)$ given by equation (4.1) only depends on the data η_0, ω_0, v_0 and $\mathbf{g}$. If we define $\hat{t}_1 := t_1 - t_*$ and $\hat{t}_2 := t_2 - t_*$

we find that both $\hat{t}_1$ and $\hat{t}_2$ must satisfy problem (4.2). For sufficiently small $0 < T \leq T_0$ and $0 < R$, this problem has a unique solution given by the fixed point of ϕ_R^T. On some time interval $J_{T'}$, where $T' \leq T$, both $\hat{t}_1$ and $\hat{t}_2$ satisfy

$$\hat{t}_i \in \mathcal{K}_R^{T'}, \quad i \in \{1, 2\},$$

as they start from initial values 0. Thus, they are fixed points of $\phi_R^{T'}$ and must coincide. This implies that t_1 and t_2 and the transforms s_1 and s_2 must also be the same on $(0, T')$. Again, this argument can be iterated to prove uniqueness on the whole maximal interval of existence.

Note that the components of q, i.e. q_0 and $\mathbf{g} \cdot Y$, depend on the choice of χ in the transform. By the above argument, however, q itself does not.

Chapter 5

Generalized Newtonian Fluids

5.1 Introduction

Newtonian fluids are characterized by the property that their shear stress depends linearly on the strain they suffer. This relation is determined by one constant parameter, their viscosity μ, i.e.

$$\mathbf{S} = \mu \mathcal{E}^{(v)},$$

where $\mathbf{S}$ denotes the fluid's deviatoric or viscous stress tensor and v its velocity. Under appropriate assumptions on the environment, water and air are examples of Newtonian fluids.

If this relation is disturbed, the fluid exhibits non-Newtonian behavior. This can happen if the material has a local structure which interacts with its flow. For example, macromolecular particles contained in a fluid can change the flow via their capacity to store energy from drag forces or via their orientation. On the other hand, the particles in Newtonian fluids are considered to be so small that their individual activity cannot disturb the flow dynamics. Physical phenomena which distinguish Newtonian from non-Newtonian fluids are for example the Weissenberg or rod-climbing effect, die swelling, i.e. the increase in diameter of flows leaving a tube, or the tubeless siphon effect. For a more detailed description and illustrations, we refer to [Ren00, Chapter 1].

Modeling of non-Newtonian fluids is a very complicated and active field of research. It is far beyond the scope of this thesis to describe its development. In the following, we give a short summary of models which appear in mathematical analysis and results related to them in order to provide a

51

background for the introduction of the special class of generalized Newtonian fluids used in our model. For surveys of both physical and mathematical aspects of these models we mainly refer to the books of Renardy, [Ren00], and Joseph, [Jos90], on viscoelastic liquids.

There are at least three parameters which may have to be included in determining the fluid stress: Recently, examples of fluids whose viscosity depends on high pressure were studied in mathematical analysis, cf. for example [BMR09] and the references therein. Secondly, temperature is a parameter whose influence is taken into account for example in first and second order fluid models. In models on viscoelastic fluids, time is an argument of the viscosity function, which means that the material has a stress memory. This is the main factor in creating the specific non-Newtonian effects described above. Roughly speaking, we can distinguish between rheopectic fluids, which increase their apparent viscosity over a time in which they experience stress, i. e. whipped cream and some lubricants, and thixopectic fluids, which decrease viscosity, like paint and tomato ketchup.

In order to model this behavior, Boltzmann's theory constitutes a relation of the integral form

$$\mathbf{S}(t,x) = 2 \int_{-\infty}^{t} G(t-s)\mathcal{E}^{(v)}(s,x)\,\mathrm{d}s$$

which includes a positive and monotonely increasing function G determing the fluid's memory. On the other hand, Maxwell's model states that the fluid stress is related to the strain via an ordinary differential equation in time, i.e.

$$\partial_t \mathbf{S} + \lambda \mathbf{S} = 2\mu\mathcal{E}^{(v)}.$$

Additional considerations and in particular the requirement that the stress tensor should be frame indifferent lead to non-linear models. For example, the *upper convected Maxwell (UCM) model* states the relation

$$\partial_t \mathbf{S} + (v \cdot \nabla)\mathbf{S} - (\nabla v)\mathbf{S} - \mathbf{S}(\nabla v)^T + \lambda \mathbf{S} = 2\mu\mathcal{E}^{(v)}.$$

There is a long list of modifications and generalizations of this theory. Again, we refer to [Jos90, Chapter 1]. Superposition of the (UCM) model with a Newtonian stress relation generates Oldroyd-B-type fluids. For this kind of model, the existence of global weak solutions was shown in [LM00] and under smallness assumptions on the data, it was extended to unbounded domains in [Sal05]. The existence of steady strong solutions was investigated in [NSV99], [AS03] and [AS05] for bounded and exterior domains.

Since in these models the stress depends non-linearly on the shear rate, in general, they may predict either a shear-thinning or a shear-thickening behavior of the fluid. This means that either the apparent viscosity decreases with growing shear as in so-called pseudoplastic materials or that it decreases, as in dilatant fluids. The first effect is much more common and it can for example be explained by the improving alignment of the particles with the flow direction or the breakdown of microstructures as stress increases. The second effect can for example be observed in concentrated solutions of starch in water. Under high stress, the fluid components are squeezed out from between the molecules and the whole material acts solid-like.

The concept of generalized Newtonian fluids is mainly constructed to capture these effects, but it is oblivious to the other factors like memory, temperature and pressure. The stress is related to the strain via a viscosity function which depends only on the symmetric part of the velocity gradient, i.e.

$$\mathbf{S} = \mu(2\mathcal{E}^{(v)}). \tag{5.1}$$

We simplify this relation to

$$\mathbf{S} := \mu(|\mathcal{E}^{(v)}|_2^2)\mathcal{E}^{(v)},$$

where

$$|\mathcal{E}^{(v)}|_2^2 = \sum_{i,j=1}^{n} (\varepsilon_{ij}^{(v)})^2$$

and $\varepsilon_{ij}^{(v)} = \frac{1}{2}(\partial_i v_j + \partial_j v_i)$. Note that due to incompressibility, the first invariant of $\mathcal{E}^{(v)}$ satisfies $\operatorname{tr}\mathcal{E}^{(v)} = \operatorname{div} v = 0$. An argument which can be used as a justification for the above simplification can be found in [MRR95, pp. 790]. Under the assumptions of incompressibility and frame indifference, (5.1) can be reduced to

$$\mathbf{S} = 2\beta_1\mathcal{E}^{(v)} + 2\beta_2(\mathcal{E}^{(v)})^2,$$

where the functions β_i depend on the second invariant

$$(\operatorname{tr}\mathcal{E}^{(v)})^2 - \operatorname{tr}(\mathcal{E}^{(v)})^2 = -\operatorname{tr}(\mathcal{E}^{(v)})^2 = -|\mathcal{E}^{(v)}|_2^2$$

of the shear rate and on $\operatorname{tr}(\mathcal{E}^{(v)})^3$. Furthermore, Málek, Rajagopal and Ružička state that the case $\beta_2 = 0$ already contains all power-law models in use. The case $\operatorname{tr}(\mathcal{E}^{(v)})^3 = 0$ occurs in plane isochoric flow and it guarantees that $\mathbf{S}$ has a potential.

Several typically non-Newtonian phenomena like the Weissenberg effect cannot be captured by this model. Instead, it is used if the interest is on global quantities like the flow rate. In mathematical analysis, this model is therefore perceived as a "starting point" (cf. [BP07, p. 380] or [Che88, pp. 226]) which is accessible more easily than others. However, it also appears in rheology literature and in applications, cf. e.g. the references in [MRR95], not only because it more readily allows for mathematical and computational treatment, but also because it is difficult to get fluid parameters from experiment, which are needed to determine more elaborate stress-strain relations.

A mathematical treatment of generalized Newtonian fluids was done first by Ladyzhenskaya, cf. [Lad69]. In this work, a polynomial relation

$$\mu(|\mathcal{E}^{(v)}|_2^2) = \mu_0 + \mu_1 |\mathcal{E}^{(v)}|_2^{d-2}$$

was assumed, where μ_0, μ_1, d are positive constants. The exponent d indicates whether a fluid is shear-thinning ($d < 2$) or shear-thickening ($d > 2$), where the range of physical interest is $d \geq 1$. If $d = 2$, the fluid is Newtonian with viscosity $\mu_0 + \mu_1$. In [MR05], fluids of this type are called *Ladyzhenskaya fluids*. The power-law-type fluids of *Ostwald and de Waele* are characterized by the relation

$$\mu(|\mathcal{E}^{(v)}|_2^2) = \mu_0 |\mathcal{E}^{(v)}|_2^{d/2-1}.$$

Here also, $\mu_0 > 0$ is a constant. If the shear rate becomes zero, this model predicts an infinite viscosity. This defect can be overcome by several modifications. As a slight generalization of Ladyzhenskaya fluids, a standard model in mathematical analysis is

$$\mu(|\mathcal{E}^{(v)}|_2^2) = \mu_0 (1 + |\mathcal{E}^{(v)}|_2^2)^{d/2-1}. \tag{5.2}$$

Starting from [Lad69], the theory of weak solutions for power-law-type fluids was developed mainly by Nečas, Málek, Růžička, Frehse and their co-authors, cf. for example [MNR93], [MRR95], [MNR01], [FMS03], [FR08] and the survey in [MR05]. The value of the exponent d plays an important role for these results. For an overview of the methods and the history of results, cf. the table in [MR05, p. 414].

In addition, there are results on the existence of strong solutions. Well-posedness for more general relations between stress and strain and under the assumption of small initial data was shown in [Ama94]. In [DR05], Diening and Růžička established the existence of unique strong solutions, locally in time for $d > \frac{7}{5}$, regardless of the smallness of initial data. In Wielage, more exponents are obtained for the equations on the whole space.

Most recently, for $p > n + 2$, Bothe and Prüss showed the existence of local L^p-strong solutions for a generalization of the model (5.2) which includes all exponents $d \geq 1$. We employ the same model in order to be able to use the linear maximal regularity result quoted in Proposition 1.7 and copy Bothe and Prüss' technique of dealing with the quasi-linear structure of the problem.

More precisely, we define $\mathbf{T}^\mu(v, q) = \mathbf{S}^\mu(v) - q\mathrm{Id}$ as the *generalized Newtonian stress tensor* of the fluid, where $\mathbf{S}^\mu(v) := \mu(|\mathcal{E}^{(v)}|_2^2)\mathcal{E}^{(v)}$ and μ is a function satisfying $\mu \in C^{1,1}(\mathbb{R}_+; \mathbb{R})$ and the conditions

$$\mu(s) > 0 \quad \text{and} \quad \mu(s) + 2s\mu'(s) > 0 \quad \text{for all } s \geq 0. \tag{5.3}$$

Clearly, these assumptions include shear-thinning $(d < 2)$ and shear-thickening $(d > 2)$ power-law-type fluids of the kind (5.2).

There is few literature regarding the mathematical analysis of the interaction of a rigid body and a non-Newtonian fluid in a situation similar to ours. In [Din07, Theorem 5.5.17] it was shown that there exists a unique local strong solution for the flow of a generalized Newtonian fluid around several obstacles which move with a prescribed rotational velocity and do not touch. The proof also relies on the result by Bothe and Prüss, so that the same fluid model is employed.

The existence of weak solutions to the coupled problem of the free movement of several obstacles in a bounded domain was treated recently in [FHN08]. It was shown that a solution exists globally in time for shear-thickening power-law-type fluids with exponent $d \geq 4$. A main aspect of this result is that the objects do not contact each other or the boundary. This effect is due to the no-slip boundary condition, which generates high stresses when objects get close even in Newtonian fluids. The assumption of a shear-thickening behavior of the fluid leads to an increase in viscosity under these circumstances.

This result also gives rise to one further issue concerning the model. We assume no-slip conditions on the interface of the solid body and the fluid. Although this is well-established for Newtonian fluids, there are examples of non-Newtonian fluids which show slip in experiments, cf. [LBS07] and the references therein. So even though we can say that the consideration of this model and the main result concerning the existence of strong solutions is a "starting point", it is not clear which kinds of fluids exhibit the appropriate properties.

The remainder of this chapter is organized in the following way. In the next section, we briefly explain notation and state the main result. In Section 5.3, we show how the generalized system changes under the coordinate transform from Chapter 2. In particular, we introduce the transformed generalized

Stokes operator. Section 5.4 deals with the linearized transformed problem in a similar way as in the Newtonian situation in Chapter 3. The last section is devoted to the proof of the main result via a contraction mapping argument, similar to Chapter 4.

5.2 Main Result

Except for the viscosity function of the fluid, the Newtonian model is not changed. Again, the rigid body occupies a bounded domain $\mathcal{B}(t)$ of class $C^{2,1}$ with boundary $\Gamma(t)$ and outer normal $n(t)$, such that its complement $\mathcal{D}(t) = \mathbb{R}^3 \setminus \overline{\mathcal{B}(t)}$ is an exterior domain filled by the fluid. The system of equations we consider on some time interval J_T, $T > 0$, includes the balance laws of inertia and angular momentum of the body and the generalized Navier-Stokes equations for the fluid, i.e.

$$\begin{cases} v_t - \operatorname{div} \mathbf{T}^\mu(v,q) + (v \cdot \nabla)v = \mathbf{g} & \text{in } J_T \times \mathcal{D}(t), \\ \operatorname{div} v = 0 & \text{in } J_T \times \mathcal{D}(t), \\ v|_{\Gamma(t)} = v_\mathcal{B} & \text{on } J_T \times \Gamma(t), \\ v(0) = v_0 & \text{in } \mathcal{D}(0), \\ \mathrm{m}\eta' + \int_{\Gamma(t)} \mathbf{T}^\mu(v,q)n \, \mathrm{d}\sigma = \mathrm{mg} & \text{in } J_T, \\ (J\omega)' + \int_{\Gamma(t)} (x - x_c) \times \mathbf{T}^\mu(v,q)n \, \mathrm{d}\sigma = 0 & \text{in } J_T, \\ \eta(0) = \eta_0, \\ \omega(0) = \omega_0. \end{cases} \quad (5.4)$$

The velocity and pressure of the fluid are v and q, whereas η and ω denote the body's translational and angular velocity. Its full velocity is given by

$$v_\mathcal{B}(t,x) := \eta(t) + \omega(t) \times (x - x_c(t)).$$

The body has mass m, density $\rho_\mathcal{B}(x)$ and inertia tensor $J(t)$. We set the constant fluid density to 1 and assume for simplicity that $x_c(0) = 0$ for the position of the body's center of gravity x_c at starting time.

As a generalization of the Navier-Stokes case where μ is a constant, we consider

$$\mathbf{T}^\mu(v,q) := 2\mu(|\mathcal{E}^{(v)}|_2^2)\mathcal{E}^{(v)} - q\mathrm{Id}, \quad (5.5)$$

where $\mu \in C^{1,1}(\mathbb{R}_+; \mathbb{R})$ satisfies the conditions in (5.3).

The following theorem is the main result of this chapter. If μ were constant, so that $\mathbf{T}^\mu = \mathbf{T}$, it covers the Newtonian situation. However, in view of Theorem 4.6, its assumptions on the integrability exponents, $p = q > 5$, are more restrictive and thus more regularity of the data is required here.

Theorem 5.1. *Assume $p > 5$, $\eta_0, \omega_0 \in \mathbb{R}^3$ and that $\mathcal{D} := \mathcal{D}(0)$ is an exterior domain of class $C^{2,1}$. Given $v_0 \in W^{2-2/p,p}(\mathcal{D})$, $\operatorname{div} v_0 = 0$ and $v_0|_\Gamma(x) = \eta_0 + \omega_0 \times x$ there exists a maximal interval J_T, $T > 0$, such that problem* (5.4) *including the stress tensor* (5.5) *satisfying* (5.3) *admits a unique strong solution*

$$
\begin{aligned}
v &\in L^p(J_T; W^{2,p}(\mathcal{D}(\cdot))) \cap W^{1,p}(J_T; L^p(\mathcal{D}(\cdot))), \\
q &= q_0 + g \cdot Y, \quad q_0 \in L^p(J_T; \widehat{W}^{1,p}(\mathcal{D}(\cdot))), \ Y \in C^1(J_T; C^\infty(\mathcal{D}(\cdot))), \\
\eta &\in W^{1,p}(J_T), \\
\omega &\in W^{1,p}(J_T).
\end{aligned}
$$

5.3 The Transformed Generalized Stokes Operator

We redefine the unknowns and data in problem (5.4) by using the coordinate transforms X, Y from Chapter 2. For $T > 0$ and $(t, y) \in [0, T] \times \mathbb{R}^3$ let

$$
\begin{aligned}
u(t, y) &:= J_Y(t, X(t, y))v(t, X(t, y)), \\
p(t, y) &:= q(t, X(t, y)), \\
\xi(t) &:= Q^T(t)\eta(t), \\
\Omega(t) &:= Q^T(t)\omega(t), \\
\mathbf{G}(t, y) &:= J_Y(t, X(t, y))\mathbf{g}, \\
T^\mu(u(t, y), p(t, y)) &:= Q^T(t)\mathbf{T}^\mu(Q(t)u(t, y), p(t, y))Q(t), \\
I &:= Q^T(t)J(t)Q(t), \\
N &:= Q^T(t)n(t), \\
\mathcal{D} := \mathcal{D}(0), \Gamma := \Gamma(0) \ &\text{and} \ \mathcal{B} := \mathcal{B}(0).
\end{aligned}
$$

The terms v_t, $(v \cdot \nabla)v$ and ∇q in (5.4) transform to $u_t + \mathcal{M}u$, $\mathcal{N}(u)$ and $\mathcal{G}p$ as in the Newtonian setting, cf. Section 2.1. The operator $\operatorname{div} \mu(|\mathcal{E}^{(v)}|_2^2)\mathcal{E}^{(v)}$ can be written as

$$
\begin{aligned}
A(v)_i := (\operatorname{div} \mu(|\mathcal{E}^{(v)}|_2^2)\mathcal{E}^{(v)})_i &= \sum_{j=1}^3 \mu(|\mathcal{E}^{(v)}|_2^2)\partial_j \varepsilon_{ij}^{(v)} + \partial_j(\mu(|\mathcal{E}^{(v)}|_2^2))\varepsilon_{ij}^{(v)} \\
&= \sum_{j=1}^3 \big[\mu(|\mathcal{E}^{(v)}|_2^2)(\partial_j^2 v_i + \partial_i \partial_j v_j) \\
&\quad + 2\mu'(|\mathcal{E}^{(v)}|_2^2)\big(\sum_{k,l=1}^3 \varepsilon_{kl}^{(v)}\partial_j \varepsilon_{kl}^{(v)}\varepsilon_{ij}^{(v)} \big) \big].
\end{aligned}
$$

Using div $v = 0$, this can be simplified to

$$
\begin{aligned}
A(v)_i &= \mu(|\mathcal{E}^{(v)}|_2^2)\Delta v_i + 2\mu'(|\mathcal{E}^{(v)}|_2^2)\sum_{j,k,l=1}^{3}\varepsilon_{ij}^{(v)}\varepsilon_{kl}^{(v)}\partial_j\partial_l v_k \\
&= \mu(|\mathcal{E}^{(v)}|_2^2)\Delta v_i + \sum_{j,k,l=1}^{3}\alpha_{ij}^{kl}(v)\partial_j\varepsilon_{kl}^{(v)},
\end{aligned}
$$

where

$$
\alpha_{ij}^{kl}(v)(t,x) := 2\mu'(|\mathcal{E}^{(v)}(t,x)|_2^2)\varepsilon_{ij}^{(v)}(t,x)\varepsilon_{kl}^{(v)}(t,x)
$$

for all $t \in J_T$, $x \in \mathcal{D}(t)$. In the following, we set

$$
\begin{aligned}
2\varepsilon_{ij}^{(v)}(t,x) &= (\partial_i v_j)(t,x) + (\partial_j v_i)(t,x) \\
&= \sum_{k,l=1}^{3}\big[(\partial_i Y_k)(t,X(t,y))(\partial_k\partial_l X_j)(t,y) \\
&\quad + (\partial_j Y_k)(t,X(t,y))(\partial_k\partial_l X_i)(t,y)\big]u_l(t,y) \\
&\quad + \big[(\partial_i Y_k)(t,X(t,y))(\partial_l X_j)(t,y) \\
&\quad + (\partial_j Y_k)(t,X(t,y))(\partial_l X_i)(t,y)\big](\partial_k u_l)(t,y) \\
&=: 2\sum_{k,l=1}^{3}\tilde{e}_{kl}^{ij}u_l(t,y) + \tilde{d}_{kl}^{ij}\partial_k u_l(t,y) \\
&=: 2\tilde{\varepsilon}_{ij}^{(u)}(t,y) \qquad\qquad\qquad\qquad\qquad (5.6)
\end{aligned}
$$

as the transformed symmetric part of the gradient of v and we use the notation $\tilde{\mathcal{E}}^{(u)} := (\tilde{\varepsilon}_{ij}^{(u)})_{ij}$. We define the quasi-linear operator which arises from the transformation of A by

$$
(\mathcal{A}(w)u)_i = \mu(|\tilde{\mathcal{E}}^{(w)}|_2^2)(\mathcal{L}u)_i + \sum_{j,k,l,m=1}^{3}a_{ij}^{klm}(w)\partial_m\tilde{\varepsilon}_{kl}^{(u)},
$$

where

$$
a_{ij}^{klm}(w)(t,y) := 2\mu'(|\tilde{\mathcal{E}}^{(w)}|_2^2)(\partial_j Y_m)(t,X(t,y))\tilde{\varepsilon}_{ij}^{(w)}(t,y)\tilde{\varepsilon}_{kl}^{(w)}(t,y).
$$

The transformed system of equations can now be written as a quasi-linear equation on a fixed domain,

$$
\begin{cases}
u_t - \mathcal{A}(u)u + \mathcal{G}p = \mathbf{G} - \mathcal{N}(u) - \mathcal{M}u & \text{in } J_T \times \mathcal{D}, \\
\operatorname{div} u = 0 & \text{in } J_T \times \mathcal{D}, \\
u(t, y) = \xi(t) + \Omega(t) \times y & t \in J_T, y \in \mathcal{D}, \\
u(0) = v_0 & \text{in } \mathcal{D}, \\
\mathrm{m}\xi' + \int_\Gamma T^\mu(u, p)N\, d\sigma = \mathrm{m}\mathbf{G}(\cdot, 0) - \mathrm{m}(\Omega \times \xi) & t \in J_T, \\
I\Omega' + \int_\Gamma y \times T^\mu(u, p)N\, d\sigma = -\Omega \times (I\Omega) & t \in J_T, \\
\xi(0) = \eta_0, \\
\Omega(0) = \omega_0,
\end{cases}
$$

(5.7)

where $\mathcal{G}, \mathcal{N}, \mathcal{M}$ and $\mathbf{G}$ are defined as in Section 2.2. It is equivalent to (5.4).

In the next section, we will consider a linearization of (5.7) and show that its solution satisfies suitable maximal regularity estimates.

5.4 Maximal Regularity of the Linearized System

In order to linearize (5.7), in particular, we have to linearize the quasi-linear operator $\mathcal{A}$ by some operator A_*. It should posses maximal L^p-regularity, so that we can transfer the idea of the proof from the Newtonian to the generalized Newtonian setting. For the transformed stress tensor T^μ which appears in the rigid body equations, we choose the non-transformed Newtonian stress tensor

$$
\mathbf{T}(u, p) := 2\mu_0 \mathcal{E}^{(u)} - \operatorname{Id} p
$$

as a linearization, where $\mu_0 := \mu(0)$. In the following, for $v \in X_{p,p}^T$, let

$$
A(v) : X_{p,p}^T \to L^p(J_T; L^p(\mathcal{D}))
$$

be given by

$$
(A(v)u)_i := \mu(|\mathcal{E}^{(v)}|_2^2)\Delta u_i + \sum_{j,k,l=1}^{3} \alpha_{ij}^{kl}(v)\partial_j \varepsilon_{kl}^{(u)}
$$

and

$$
\alpha_{ij}^{kl}(v)(t, y) := 2\mu'(|\mathcal{E}^{(v)}|_2^2)\varepsilon_{ij}^{(v)}(t, y)\varepsilon_{kl}^{(v)}(t, y).
$$

In particular, we freeze $A(v)$ to

$$
A_* u := A(u^*)u,
$$

where $u^* \in X_{p,p}^T$ and p^*, ξ^*, Ω^* are reference solutions of the problem

$$
\begin{cases}
u_t^* - \Delta u^* + \nabla(p^* - \mathbf{g} \cdot y) = 0 & \text{in } J_T \times \mathcal{D}, \\
\operatorname{div} u^* = 0 & \text{in } J_T \times \mathcal{D}, \\
u^*(t, y) - \xi^*(t) - \Omega^*(t) \times y = 0, & t \in J_T, y \in \Gamma, \\
u^*(0) = v_0 & \text{in } \mathcal{D}, \\
\mathrm{m}(\xi^*)' + \int_\Gamma \mathbf{T}(u^*, p^*) N \, d\sigma = \mathrm{mg} & \text{in } J_T, \\
I(\Omega^*)' + \int_\Gamma y \times (\mathbf{T}(u^*, p^*) N) \, d\sigma = 0 & \text{in } J_T, \\
\xi^*(0) = \eta_0, \\
\Omega^*(0) = \omega_0.
\end{cases}
\tag{5.8}
$$

The existence of a solution to this problem follows from Theorem 3.1.

We define $\hat{u} := u - u^*, \hat{p} := p - p^* - \mathbf{g} \cdot y, \hat{\xi} := \xi - \xi^*, \hat{\Omega} := \Omega - \Omega^*$ and add $A_* \hat{u}$ to the first line of (5.7). Setting

$$
\mathcal{Q}(u^*, \hat{u}) := A_* \hat{u} - \mathcal{A}(u^* + \hat{u})(u^* + \hat{u})
$$

and

$$
\begin{aligned}
G_0(\hat{u}, \hat{p}, \hat{\xi}, \hat{\Omega}) :=\ & \mathbf{G} - \mathbf{g} - \mathcal{I}(\hat{u} + u^*, \hat{p} + p^*, \hat{\xi} + \xi^*, \hat{\Omega} + \Omega^*) - \mathcal{Q}(u^*, \hat{u}) - \Delta u^*, \\
\mathcal{I}(u, p, \xi, \Omega) :=\ & \mathcal{M}u + (\mathcal{G} - \nabla)p - \mathcal{N}(u) \\
G_1(\hat{u}, \hat{p}, \hat{\xi}, \hat{\Omega}) :=\ & \int_\Gamma (\mathbf{T} - T^\mu)(\hat{u}, \hat{p}) N \, d\sigma + \int_\Gamma (\mathbf{T} - T^\mu)(u^*, p^*) N \, d\sigma \\
& \mathrm{m}(\mathbf{G}(\cdot, 0) - \mathbf{g}) - (\Omega^* + \hat{\Omega}) \times (\xi^* + \hat{\xi}) \\
G_2(\hat{u}, \hat{p}, \hat{\xi}, \hat{\Omega}) :=\ & \int_\Gamma y \times (\mathbf{T} - T^\mu)(\hat{u}, \hat{p}) N \, d\sigma + \int_\Gamma y \times (\mathbf{T} - T^\mu)(u^*, p^*) N \, d\sigma \\
& - (\Omega^* + \hat{\Omega}) \times I(\Omega^* + \hat{\Omega}),
\end{aligned}
\tag{5.9}
$$

the system

$$
\begin{cases}
\hat{u}_t - A_* \hat{u} + \nabla \hat{p} = G_0(\hat{u}, \hat{p}, \hat{\xi}, \hat{\Omega}) & \text{in } J_T \times \mathcal{D}, \\
\operatorname{div} \hat{u} = 0 & \text{in } J_T \times \mathcal{D}, \\
\hat{u}(t, y) = \hat{\xi}(t) + \hat{\Omega}(t) \times y, & t \in J_T, y \in \Gamma, \\
\hat{u}(0) = 0 & \text{in } \mathcal{D}, \\
\int_\Gamma \mathbf{T}(\hat{u}, \hat{p}) N \, d\sigma + \mathrm{m}\hat{\xi}' = G_1(\hat{u}, \hat{p}, \hat{\xi}, \hat{\Omega}), & \text{in } J_T, \\
\int_\Gamma y \times \mathbf{T}(\hat{u}, \hat{p}) N \, d\sigma + I\hat{\Omega}' = G_2(\hat{u}, \hat{p}, \hat{\xi}, \hat{\Omega}), & \text{in } J_T, \\
\hat{\xi}(0) = 0, \\
\hat{\Omega}(0) = 0,
\end{cases}
\tag{5.10}
$$

is equivalent to (5.7).

Fixing G_0, G_1, G_2 yields the linearization of (5.7) we want to consider in this section. The main result is the following.

Theorem 5.2. *Let $\mathcal{D}$ be an exterior domain of class $C^{2,1}$, $T_0 > 0$ and $p > 5$. Assume that $g_0 \in L^p(J_{T_0}; L^p(\mathcal{D})), g_1, g_2 \in L^p(J_{T_0})$. Then the problem*

$$\begin{cases} u_t - A_* u + \nabla p &= g_0 \quad \text{in } J_{T_0} \times \mathcal{D}, \\ \operatorname{div} u &= 0 \quad \text{in } J_{T_0} \times \mathcal{D}, \\ u(t, y) - \xi(t) - \Omega(t) \times y &= 0, \quad t \in J_{T_0}, y \in \Gamma, \\ \int_\Gamma \mathbf{T}(u, p) N \, d\sigma + \mathrm{m}\xi' &= g_1 \quad \text{in } J_{T_0}, \\ \int_\Gamma y \times (\mathbf{T}(u, p) N) \, d\sigma + I\Omega' &= g_2 \quad \text{in } J_{T_0}, \end{cases} \quad (5.11)$$

with initial conditions

$$u(0) = 0, \xi(0) = \Omega(0) = 0,$$

admits a unique solution

$$u \in X^{T_0}_{p,p,0}, \quad p \in Y^{T_0}_{p,p}, \quad (\xi, \Omega) \in {}_0W^{1,p}(J_{T_0}; \mathbb{R}^6)$$

which satisfies the estimate

$$\|u\|_{X^{T_0}_{p,p}} + \|p\|_{Y^{T_0}_{p,p}} + \|(\xi, \Omega)\|_{W^{1,p}(J_{T_0})} \leq C(\|g_0\|_{p,p} + \|(g_1, g_2)\|_{L^p(J_{T_0})}), \quad (5.12)$$

where the constant C does not depend on g_0, g_1 or g_2.

As in the Newtonian case, it will again be important to rewrite the tangential component of the inhomogeneous boundary data as an external force on the fluid in $L^p_\sigma(\mathcal{D})$. In Lemma 5.3, we show that this still yields improved estimates on the pressure in the generalized setting. In Subsection 5.3.2, we give a reformulation of (5.11) in $\hat{\xi}, \hat{\Omega}$, analogous to equation (3.15) in Chapter 3. In essence, we can then repeat the arguments from Section 3.2 to show the estimates which prove the solvability of this reformulation for sufficiently small T and thus show Theorem 5.2.

5.4.1 Estimates for the Fluid Part

In [BP07, Section 3], it is shown that the operator $-A_*$ is strongly normally elliptic and that its coefficients satisfy the regularity assumptions needed in Proposition 1.7, provided we have the embedding

$$X^T_{p,p} \hookrightarrow C(J_T; W^{2-2/p,p}(\mathcal{D})) \hookrightarrow C([0,T]; C^1(\overline{\mathcal{D}})). \quad (5.13)$$

This condition leads to the assumption $p > 5$. It follows that the problem

$$\begin{cases} u_t - A_* u + \nabla p &= f \quad \text{in } J_T \times \mathcal{D}, \\ \operatorname{div} u &= 0 \quad \text{in } J_T \times \mathcal{D}, \\ u|_\Gamma &= h \quad \text{on } J_T \times \Gamma, \\ u(0) &= u_0 \quad \text{in } \mathcal{D}, \end{cases}$$

has a unique strong solution $\mathcal{U}_{A_*}(f, h, u_0), \mathcal{P}_{A_*}(f, h, u_0)$ satisfying

$$
\begin{aligned}
&\|\mathcal{U}_{A_*}(f, h, u_0)\|_{X_{p,p}^T} + \|\mathcal{P}_{A_*}(f, h, u_0)\|_{Y_{p,p}^T} \\
&\leq \ C(\|f\|_{p,p} + \|h\|_{W_p^{T,\Gamma}} + \|u_0\|_{W^{2-2/p,p}(\mathcal{D})}),
\end{aligned} \tag{5.14}
$$

where $W_p^{T,\Gamma} = W^{1-1/(2p),p}(J_T; L^p(\Gamma)) \cap L^p(J_T; W^{2-1/p,p}(\Gamma))$. Moreover, we need the following estimate on the pressure, which corresponds to Lemma 1.10 in the Newtonian situation.

Lemma 5.3. *Let* $u := \mathcal{U}_{A_*}(f, 0, 0) \in X_{p,p}^T$ *and* $\mathcal{P}_{A_*}(f, 0, 0) \in Y_{p,p}^T$ *solutions of*

$$
\left\{
\begin{aligned}
u_t - A_* u + \nabla p &= f & &\text{in } J_T \times \mathcal{D}, \\
\operatorname{div} u &= 0 & &\text{in } J_T \times \mathcal{D}, \\
u|_\Gamma &= 0 & &\text{on } J_T \times \Gamma, \\
u(0) &= 0 & &\text{in } \mathcal{D},
\end{aligned}
\right. \tag{5.15}
$$

where $f \in L^p(J_T; L^p_\sigma(\Omega))$ *and* $p > 5$. *Choose* $R > 0$, $\mathcal{D}_R = \mathcal{D} \cap B_R$ *and* $p_R = \mathcal{P}_{A_*}(f, 0, 0)$ *from* $L^p(J_T; \widehat{W}^{1,p}(\mathcal{D}))$ *such that* $p_R \in L^p(J_T; L^p_0(\mathcal{D}_R))$. *Then*

$$
\|p_R\|_{L^p(J_T; L^p(\mathcal{D}_R))} \ \leq \ CT^{\alpha/p}\|\mathcal{U}_{A_*}(f, 0, 0)\|_{X_{p,p}^T}
$$

for $\alpha = \frac{1}{2} - \frac{1}{2p} - \frac{\varepsilon}{2}$ *and* $0 < \varepsilon < 1 - \frac{1}{p}$.

Proof. Since u, p_R strongly solve (5.15) and $f \in L^q_\sigma(\mathcal{D})$, it follows that

$$
\nabla p_R(t, x) = ((\operatorname{Id} - P_{\mathcal{D},p})(A_* u)(x))(t) \quad \text{for a. a. } t \in J_T. \tag{5.16}
$$

Solving the following Neumann problem, we construct suitable test functions for p_R. Let $\psi \in L^{p'}_0(\mathcal{D}_R) = (L^p_0(\mathcal{D}_R))'$ and extend ψ to $\mathcal{D} \backslash \mathcal{D}_R$ by 0. Then by [SS07, Prop. 5.6], there is a solution ϕ_ψ of

$$
\left\{
\begin{aligned}
\Delta \phi_\psi &= \psi & &\text{in } \mathcal{D}, \\
\frac{\partial \phi_\psi}{\partial N} &= 0 & &\text{on } \Gamma,
\end{aligned}
\right.
$$

which satisfies the estimate

$$
\|\nabla \phi_\psi\|_{W^{1,p'}(\mathcal{D})} \ \leq \ C\|\psi\|_{L^{p'}(\mathcal{D})}.
$$

Integrating by parts and using (5.16) and $\int_\Gamma p_R(\nabla \phi_\psi \cdot N) = 0$, it follows that

$$\int_{\mathcal{D}_R} p_R(t)\psi$$

$$= \int_{\mathcal{D}} p_R(t)\Delta \phi_\psi$$

$$= -\int_{\mathcal{D}} \nabla p_R(t) \cdot \nabla \phi_\psi + \int_\Gamma p_R(t)(\nabla \phi_\psi \cdot N)$$

$$= -\int_{\mathcal{D}} (\mathrm{Id} - P_{\mathcal{D},p}) A_* u \cdot \nabla \phi_\psi$$

$$= -\sum_{i=1}^3 \int_{\mathcal{D}} \Big[\mu(|\mathcal{E}^{(u^*(t))}|_2^2)\Delta u_i(t) + \sum_{j,k,l=1}^3 \alpha_{ij}^{kl}\partial_j\partial_k u_l(t) \Big] \big[(\mathrm{Id} - P_{\mathcal{D},p})\nabla \phi_\psi \big]_i$$

$$= -\sum_{i=1}^3 \Big[\int_{\mathcal{D}} \mu(|\mathcal{E}^{(u^*(t))}|_2^2)\Delta u_i(t)\partial_i \phi_\psi + \sum_{j,k,l=1}^3 \int_{\mathcal{D}} \partial_j\partial_k u_l(t)\alpha_{ij}^{kl}\partial_i \phi_\psi \Big]$$

for almost all $t \in J_T$. We use integration by parts a second time to write

$$\sum_{i=1}^3 \int_{\mathcal{D}} \mu(|\mathcal{E}^{(u^*(t))}|_2^2)\Delta u_i(t)\partial_i \phi_\psi = \sum_{i=1}^3 \Big[-\int_{\mathcal{D}} \nabla u_i(t) \cdot \nabla \big(\mu(|\mathcal{E}^{(u^*(t))}|_2^2)\partial_i \phi_\psi \big)$$
$$+ \int_\Gamma \mu(|\mathcal{E}^{(u^*(t))}|_2^2)\partial_i \phi_\psi(\nabla u_i(t) \cdot N) \Big]$$
$$=: \ I + II,$$

as well as

$$\sum_{i,j,k,l=1}^3 \int_{\mathcal{D}} \partial_j\partial_k u_l(t)\alpha_{ij}^{kl}\partial_i \phi_\psi = \sum_{i,j,k,l=1}^3 \Big[-\int_{\mathcal{D}} \partial_k u_l(t)\partial_j(\alpha_{ij}^{kl}\partial_i \phi_\psi)$$
$$+ \int_\Gamma \partial_k u_l(t)(\alpha_{ij}^{kl}\partial_i \phi_\psi N_j) \Big]$$
$$=: \ I_* + II_*.$$

Let $s = \frac{p}{p-2}$ to get $\frac{1}{s} + \frac{2}{p} = 1$. By Hölder's inequality and by Proposition 1.1 we get

$$|I| \le \|\nabla u(t)\|_{L^p(\mathcal{D})}\|\nabla(\mu(|\mathcal{E}^{(u^*(t))}|_2^2))\|_{L^p(\mathcal{D})}\|\nabla \phi_\psi\|_{L^s(\mathcal{D})}$$
$$+ \|\nabla u(t)\|_{L^p(\mathcal{D})}\|\mu(|\mathcal{E}^{(u^*(t))}|_2^2)\|_{L^\infty(\mathcal{D})}\|\nabla^2 \phi_\psi\|_{L^{p'}(\mathcal{D})},$$

$$|I_*| \le \sup_{i,j,k,l}\big(\|\nabla u(t)\|_{L^p(\mathcal{D})}\|\partial_j\alpha_{ij}^{kl}\|_{L^p(\mathcal{D})}\|\nabla \phi_\psi\|_{L^s(\mathcal{D})}$$
$$+ \|\nabla u(t)\|_{L^p(\mathcal{D})}\|\alpha_{ij}^{kl}\|_{L^\infty(\mathcal{D})}\|\nabla^2 \phi_\psi\|_{L^{p'}(\mathcal{D})}\big),$$

$$\begin{aligned}
|II| &\leq \|\nabla u(t)\|_{L^p(\Gamma)}\|\mu(|\mathcal{E}^{(u^*)}|^2_2)\|_{L^\infty(\Gamma)}\|\nabla\phi_\psi\|_{L^{p'}(\Gamma)} \\
&\leq C\|u(t)\|_{W^{1+1/p+\varepsilon,p}(\mathcal{D})}\|\mu(|\mathcal{E}^{(u^*)}|^2_2)\|_{C(\overline{\mathcal{D}})}\|\nabla\phi_\psi\|_{W^{1,p'}(\mathcal{D})}, \\
|II_*| &\leq \sup_{i,j,k,l}\|\nabla u(t)\|_{L^p(\Gamma)}\|\alpha_{ij}^{kl}\|_{L^\infty(\Gamma)}\|\nabla\phi_\psi\|_{L^{p'}(\Gamma)} \\
&\leq C\|u(t)\|_{W^{1+1/p+\varepsilon,p}(\mathcal{D})}\|\alpha_{ij}^{kl}\|_{C(\overline{\mathcal{D}})}\|\nabla\phi_\psi\|_{W^{1,p'}(\mathcal{D})}
\end{aligned}$$

for $0 < \varepsilon < 1 - 1/p$. From Sobolev embeddings in $\mathbb{R}^3$ it follows that $W^{1,p'}(\mathcal{D}) \hookrightarrow L^s(\mathcal{D})$. The embedding (5.13) implies

$$\|\nabla(\mu(|\mathcal{E}^{(u^*)}|^2_2))\|_{L^p(\mathcal{D})} + \|\mu(|\mathcal{E}^{(u^*)}|^2_2)\|_{C(\overline{\mathcal{D}})} + \|\partial_j\alpha_{ij}^{kl}\|_{L^p(\mathcal{D})} + \|\alpha_{ij}^{kl}\|_{C(\overline{\mathcal{D}})} \leq C_0$$

for all i, j, k, l and almost all t. Thus

$$|I| + |I_*| \leq C\|\nabla u(t)\|_{L^p(\mathcal{D})}\|\nabla\phi_\psi\|_{W^{1,p'}(\mathcal{D})} \leq C\|\nabla u(t)\|_{L^p(\mathcal{D})}\|\psi\|_{L^{p'}(\mathcal{D})}$$

and

$$\begin{aligned}
|II| + |II_*| &\leq C\|u(t)\|_{W^{1+1/p+\varepsilon,p}(\mathcal{D})}\|\nabla\phi_\psi\|_{W^{1,p'}(\mathcal{D})} \\
&\leq C\|u(t)\|_{W^{1+1/p+\varepsilon,p}(\mathcal{D})}\|\psi\|_{L^{p'}(\mathcal{D})}.
\end{aligned}$$

Setting $\alpha = \frac{1}{2} - \frac{1}{2p} - \frac{\varepsilon}{2} > 0$ implies

$$\|p_R(t)\|_{L^p(\mathcal{D}_R)} \leq C\|u(t)\|_{W^{2-2\alpha,p}(\mathcal{D})}$$

for almost all $t \in J_T$. By interpolation and by Proposition 1.8 it follows that

$$\begin{aligned}
\|p_R\|_{L^p(J_T;L_0^p(\mathcal{D}_R))} &\leq C\|u\|_{L^p(J_T;W^{2,p}(\mathcal{D}))}^{1-\alpha}\|u\|_{p,p}^\alpha \\
&\leq CT^{\alpha/p}\|u\|_{X_{p,p}^T}^{1-\alpha}\|u\|_{L^\infty(J_T;L^p(\mathcal{D}))}^\alpha \\
&\leq CT^{\alpha/p}\|u\|_{X_{p,p}^T} \\
&\leq CT^{\alpha/p}\|f\|_{p,p},
\end{aligned}$$

where the constant C may be chosen independently of T. $\qquad\qquad\square$

5.4.2   Proof of Theorem 5.2

The estimates from the last subsection allow us to repeat the reformulation of the linearized problem and the subsequent estimates from Chapter 3 in a similar way. First we correct for the normal component of the boundary using the function $v_{\xi,\Omega}$ defined in (3.7). For almost all $t \in J_T$, $0 < T < T_0$, it satisfies the Neumann problem

$$\begin{cases}
\Delta v_{\xi,\Omega}(t) = 0 & \text{in } \mathcal{D}, \\
\frac{\partial v_{\xi,\Omega}(t)}{\partial N}|_\Gamma(y) = (\xi(t) + \Omega(t) \times y) \cdot N(y), & y \in \Gamma,
\end{cases}$$

and the estimate

$$\|\nabla v_{\xi,\Omega}\|_{W^{1,p}(J_T;W^{2,p}(\mathcal{D}))} + \|\partial_t v_{\xi,\Omega}\|_{Y^T_{p,p}} \leq C\|(\xi,\Omega)\|_{W^{1,p}(J_T)}.$$

We define the momentum matrix $\mathbb{I}$, the added mass $\mathbb{M}$ and the operator $\mathcal{J}$ as before in (3.10), (3.12) and (3.11). This yields the equation

$$\begin{pmatrix} \hat{\xi} \\ \hat{\Omega} \end{pmatrix} = \mathcal{R}_{A_*} \begin{pmatrix} \hat{\xi} \\ \hat{\Omega} \end{pmatrix} + g^*, \tag{5.17}$$

as an equivalent reformulation of (5.11), where

$$\mathcal{R}_{A_*} : {}_0W^{1,p}(J_T;\mathbb{R}^6) \to {}_0W^{1,p}(J_T;\mathbb{R}^6)$$

is given by

$$\mathcal{R}_{A_*}(\hat{\xi},\hat{\Omega})(t) := \int_0^t (\mathbb{I}+\mathbb{M})^{-1} \mathcal{J}\left[\mathbf{T}(\mathcal{U}_{h,A_*}(\hat{\xi},\hat{\Omega}), \mathcal{P}_{h,A_*}(\hat{\xi},\hat{\Omega}))\right](s)\,\mathrm{d}s$$

and

$$g^*(t) := \int_0^t (\mathbb{I}+\mathbb{M})^{-1}\left[\begin{pmatrix} g_1 \\ g_2 \end{pmatrix} + \mathcal{J}\mathbf{T}(\mathcal{U}_{A_*}(g_0,0,0), \mathcal{P}_{A_*}(g_0,0,0))\right]\mathrm{d}s.$$

We only have to modify the argument for the Newtonian case by replacing $\mathcal{U}_h(\hat{\xi},\hat{\Omega}), \mathcal{P}_h(\hat{\xi},\hat{\Omega})$ with $\mathcal{U}_{h,A_*}(\hat{\xi},\hat{\Omega}) := \mathcal{U}_{A_*}(0,\hat{\xi}+\hat{\Omega}\times y - \nabla v_{\hat{\xi},\hat{\Omega}}|_\Gamma, 0)$ and $\mathcal{P}_{h,A_*}(\hat{\xi},\hat{\Omega}) := \mathcal{P}_{A_*}(0,\hat{\xi}+\hat{\Omega}\times y - \nabla v_{\hat{\xi},\hat{\Omega}}|_\Gamma, 0)$. From (5.14) and the properties of $v_{\hat{\xi},\hat{\Omega}}$, it follows that

$$\|\mathcal{U}_{h,A_*}(\hat{\xi},\hat{\Omega})\|_{X^T_{p,p}} + \|\mathcal{P}_{h,A_*}(\hat{\xi},\hat{\Omega})\|_{Y^T_{p,p}} \leq C\|\hat{\xi}+\hat{\Omega}\times\cdot - \nabla v_{\hat{\xi},\hat{\Omega}}\|_{W^{T,\Gamma}_p}$$
$$\leq C\|(\hat{\xi},\hat{\Omega})\|_{W^{1,p}(J_T)}.$$

Repeating the estimates on $\mathcal{J}$ and using that

$$\|\mathcal{P}_{h,A_*}(\hat{\xi},\hat{\Omega})\|_{L^p(J_T;L^q_0(\mathcal{D}_R))} \leq CT^{\alpha/p}\|\mathcal{U}_{h,A_*}(\hat{\xi},\hat{\Omega})\|_{X^T_{p,q}}$$
$$\leq CT^{\alpha/p}\|(\hat{\xi},\hat{\Omega})\|_{W^{1,p}(J_T)}$$

holds for $\alpha = \frac{1}{2} - \frac{1}{2p} - \frac{\varepsilon}{2}$, $0 < \varepsilon < 1 - \frac{1}{p}$ thanks to Lemma 5.3, we see that

$$\|\mathcal{R}_{A_*}(\hat{\xi},\hat{\Omega})\|_{W^{1,p}(0,T)} \leq C(T + T^{1/3p} + T^{c\alpha/p})\|(\hat{\xi},\hat{\Omega})\|_{W^{1,p}(J_T)}$$

and clearly,

$$\|g^*\|_{W^{1,p}(0,T)} \leq C\left\|\begin{pmatrix} g_1 \\ g_2 \end{pmatrix} + \mathcal{J}\mathbf{T}(\mathcal{U}_{A_*}(g_0,0,0), \mathcal{P}_{A_*}(g_0,0,0))\right\|_p$$
$$\leq C(\|g_0\|_{p,p} + \|g_1\|_p + \|g_2\|_p).$$

For sufficiently small $T > 0$, we get $\|\mathcal{R}_{A_*}\|_{L(_0W^{1,p}(J_T))} < 1$, which yields a solution $(\hat{\xi}, \hat{\Omega})$ of (5.17). The fluid velocity $\hat{u}$ and pressure $\hat{p}$ are given by

$$\hat{u} = \mathcal{U}_{h,A_*}(\hat{\xi}, \hat{\Omega}) + \mathcal{U}_{A_*}(g_0, 0, 0) + \nabla v_{\hat{\xi}, \hat{\Omega}}$$

and

$$\hat{p} = \mathcal{P}_{h,A_*}(\hat{\xi}, \hat{\Omega}) + \mathcal{P}_{A_*}(g_0, 0, 0) - \partial_t v_{\hat{\xi}, \hat{\Omega}}.$$

Thus, if we repeat the arguments in the proof of Theorem 3.1 to extend this solution from J_T to J_{T_0}, Theorem 5.2 is proved.

5.5   The Fixed Point Argument and Estimates for the Additional Non-Linearities

As in Chapter 4, we solve (5.10) by using Theorem 5.2 and a contraction mapping argument. For $p > 5$, we choose

$$\mathcal{K}_R^T := \{(\tilde{u}, \tilde{p}, \tilde{\xi}, \tilde{\Omega}) \in X_{p,p,0}^T \times Y_{p,p}^T \times {}_0W^{1,p}(J_T; \mathbb{R}^6) :$$
$$\|\tilde{u}\|_{X_{p,p}^T} + \|\tilde{p}\|_{Y_{p,p}^T} + \|(\tilde{\xi}, \tilde{\Omega})\|_{W^{1,p}(J_T)} \leq R\}$$

as the closed ball on which to construct a map ψ_R^T which has a unique fixed point that solves (5.10).

Let G_0, G_1 and G_2 be the right hand sides of (5.10) as in (5.9). Let

$$\psi_R^T : \begin{pmatrix} \tilde{u} \\ \tilde{p} \\ \tilde{\xi} \\ \tilde{\Omega} \end{pmatrix} \mapsto \begin{pmatrix} G_0(\tilde{u}, \tilde{p}, \tilde{\xi}, \tilde{\Omega}) \\ G_1(\tilde{u}, \tilde{p}, \tilde{\xi}, \tilde{\Omega}) \\ G_2(\tilde{u}, \tilde{p}, \tilde{\xi}, \tilde{\Omega}) \end{pmatrix} \overset{\text{Thm. 5.2}}{\mapsto} \begin{pmatrix} u \\ p \\ \xi \\ \Omega \end{pmatrix},$$

the function which first maps $\tilde{u}, \tilde{p}, \tilde{\xi}, \tilde{\Omega}$ to G_0, G_1, G_2 and then to the solution of the linear problem with fixed right hand sides, using Theorem 5.2. For sufficiently small $R, T > 0$, we show that the Banach fixed point theorem can be applied to ψ_R^T.

For the remainder of the chapter, we assume $(\tilde{u}, \tilde{p}, \tilde{\xi}, \tilde{\Omega}) \in \mathcal{K}_R^T$ and that $u^*, p^*, \xi^*, \Omega^*$ are given by (5.8). We set

$$C_0 := \|u^*\|_{X_{p,p}^{T_0}} + \|p^* - g \cdot y\|_{Y_{p,p}^{T_0}} + \|\xi^*\|_{W^{1,p}(J_{T_0})} + \|\Omega^*\|_{W^{1,p}(J_{T_0})}$$

and let $K_*(T) := \|u^*\|_{X_{p,p}^T}$, so that $K_*(T) \to 0$ as $T \to 0$. Note that the embedding constants in (5.13) are uniform in T on the subspace $X_{p,p,0}^T$ but they blow up when $T \to 0$ in general. Let now $C_0^R := (C_0 + R)^2$. Since $\mu \in C^{1,1}(\mathbb{R}_+)$, we may define

$$m_\mu := \|\mu\|_{C^1([0,C_0^R])} + \sup_{x,y \in \mathbb{R}_+, x \neq y} \frac{|\mu'(x) - \mu'(y)|}{|x - y|}.$$

This definition will make sense as in the following estimates, we insert the arguments $|\mathcal{E}^{(u^* + \tilde{u})}|_2^2 \leq C_0^R$ into μ.

We need estimates of the type

$$\|\psi_R^T(\tilde{u}, \tilde{p}, \tilde{\xi}, \tilde{\Omega})\|_{\mathcal{K}_R^T} \leq L(T, R)\|(\tilde{u}, \tilde{p}, \tilde{\xi}, \tilde{\Omega})\|_{\mathcal{K}_R^T},$$

where $L(T, R) \to 0$ as $T, R \to 0$. Thanks to maximal regularity of the linear problem, they follow directly from estimates like

$$\|G_0(\tilde{u}, \tilde{p}, \tilde{\xi}, \tilde{\Omega})\|_{p,p} \leq \frac{1}{3}L(T, R)\|(\tilde{u}, \tilde{p}, \tilde{\xi}, \tilde{\Omega})\|_{\mathcal{K}_R^T}$$

for the functions G_0, G_1, G_2.

The proof requires several estimates known from the Newtonian case, in particular, Lemma 4.2, where we dealt with the functions $\mathbf{G} - \mathbf{g}$ and $\mathcal{I}$ in G_0. We show the estimate for the remaining term

$$\mathcal{Q}(u^*, \tilde{u}) = A_*\tilde{u} - \mathcal{A}(\tilde{u} + u^*)(\tilde{u} + u^*).$$

It depends more strongly on the structure of the linearization A_* than the terms from the Newtonian situation, so that a lot of terms have to be sorted in a convenient way. Apart from this work, the main ingredients are, as in Chapter 4, the Hölder inequality and embeddings. We define

$$\begin{aligned}
A_*\tilde{u} - \mathcal{A}(\tilde{u} + u^*)(\tilde{u} + u^*) &= (A_* - \mathcal{A}(\tilde{u} + u^*))(\tilde{u} + u^*) - A_*u^* \\
&\quad + [A(u^* + \tilde{u}) - \mathcal{A}(\tilde{u} + u^*)](\tilde{u} + u^*) \\
&=: \mathcal{Q}_I + \mathcal{Q}_{II} + \mathcal{Q}_{III},
\end{aligned}$$

so that $\mathcal{Q}_I$ refers to the difference between the frozen and the quasi-linear generalized Stokes operator and $\mathcal{Q}_{III}$ is given by the difference between the original and the transformed generalized Stokes operator. First we define $w := \tilde{u} + u^*$ for simplicity and keep in mind that

$$\|w\|_{X_{p,p}^T} \leq R + K_*(T) \quad \text{and} \quad \|w\|_{\infty,\infty} + \|\mathcal{E}^{(w)}\|_{\infty,\infty} \leq C(R + C_0)$$

due to the embedding (5.13). In the following, we deal with $\mathcal{Q}_I$ as in [BP07, p. 417]. By definition,

$$
\begin{aligned}
((A_* - A(w))w)_i &= (\mu(|\mathcal{E}^{(u^*)}|_2^2) - \mu(|\mathcal{E}^{(w)}|_2^2))\Delta w_i \\
&\quad + \sum_{j,k,l=1}^{3} [\alpha_{ij}^{kl}(u^*) - \alpha_{ij}^{kl}(w)]\partial_j\partial_k w_l
\end{aligned}
$$

and

$$
\begin{aligned}
&\alpha_{ij}^{kl}(u^*) - \alpha_{ij}^{kl}(u^* + \tilde{u}) \\
&= 4\left(\mu'(|\mathcal{E}^{(u^*)}|_2^2)\varepsilon_{ij}^{(u^*)}\varepsilon_{kl}^{(u^*)} - \mu'(|\mathcal{E}^{(u^*+\tilde{u})}|_2^2)\varepsilon_{ij}^{(u^*+\tilde{u})}\varepsilon_{kl}^{(u^*+\tilde{u})}\right) \\
&= 4\left(\mu'(|\mathcal{E}^{(u^*)}|_2^2) - \mu'(|\mathcal{E}^{(u^*+\tilde{u})}|_2^2)\right)\varepsilon_{ij}^{(u^*)}\varepsilon_{kl}^{(u^*)} \\
&\quad + \mu'(|\mathcal{E}^{(u^*+\tilde{u})}|_2^2)(\varepsilon_{ij}^{(u^*)}\varepsilon_{kl}^{(u^*)} - \varepsilon_{ij}^{(u^*+\tilde{u})}\varepsilon_{kl}^{(u^*+\tilde{u})}).
\end{aligned}
$$

We calculate

$$
\begin{aligned}
\|\mu'(|\mathcal{E}^{(u^*)}|_2^2) - \mu'(|\mathcal{E}^{(u^*+\tilde{u})}|_2^2)\|_{\infty,\infty} &\leq m_\mu \sum_{i,j=1}^{3} \|(\varepsilon_{ij}^{(u^*)})^2 - (\varepsilon_{ij}^{(u^*)} + \varepsilon_{ij}^{(u^*+\tilde{u})})^2\|_{\infty,\infty} \\
&\leq C \sup_{i,j} \|\varepsilon_{ij}^{(\tilde{u})}(\varepsilon_{ij}^{(\tilde{u})} - 2\varepsilon_{ij}^{(u^*)})\|_{\infty,\infty} \\
&\leq CR(R + C_0)
\end{aligned}
$$

and

$$
\begin{aligned}
&\|\alpha_{ij}^{kl}(u^*) - \alpha_{ij}^{kl}(u^* + \tilde{u})\|_{\infty,\infty} \\
&\leq CR(R + C_0) + Cm_\mu \sup_{i,j,k,l} \|\varepsilon_{ij}^{(u^*)}\varepsilon_{kl}^{(\tilde{u})} + \varepsilon_{ij}^{(\tilde{u})}\varepsilon_{kl}^{(u^*+\tilde{u})}\|_{\infty,\infty} \\
&\leq CR(R + C_0),
\end{aligned}
$$

so that

$$
\|\mathcal{Q}_I\|_{p,p} \leq CR(R + C_0)(\|\Delta w\|_{p,p} + \|D^2 w\|_{p,p}) \leq CR(R + C_0)(R + K_*(T)).
$$

Moreover, it follows immediately from the definition that $\|\mathcal{Q}_{II}\|_{p,p} \leq CK_*(T)$. We "add zeros" in the last term $\mathcal{Q}_{III}$ to define

$$
\begin{aligned}
[\mathcal{Q}_{III}]_i &= [(A(w) - \mathcal{A}(w))w]_i \\
&= \left(\mu(|\tilde{\mathcal{E}}^{(w)}|_2^2)(\Delta - \mathcal{L})w_i\right) + \left(\mu(|\mathcal{E}^{(w)}|_2^2) - \mu(|\tilde{\mathcal{E}}^{(w)}|_2^2)\right)\Delta w_i \\
&\quad + \left(\sum_{j,k,l=1}^{3} \alpha_{ij}^{kl}(w)\partial_j\partial_k w_l - \sum_{j,k,l,m=1}^{3} a_{ij}^{klm}(w)\partial_m\tilde{\varepsilon}_{kl}^{(w)}\right) \\
&=: (i) + (ii) + (iii).
\end{aligned}
$$

From Lemma 4.2, it follows that

$$\|(i)\|_{p,p} \leq Cm_\mu\|(\Delta - \mathcal{L})w\|_{p,p} \leq CT\|w\|_{X_{p,p}^T} \leq CT(R + K_*(T)). \quad (5.18)$$

For brevity, we omit the argument w in $\varepsilon^{(w)}, \tilde{\varepsilon}^{(w)}, \alpha_{ij}^{kl}(w), \ldots$ in the following estimates. The second term satisfies

$$
\begin{aligned}
\|(ii)\|_{p,p} &\leq Cm_\mu \sum_{i,j=1}^{3} \|\varepsilon_{ij}^2 - \tilde{\varepsilon}_{ij}^2\|_{\infty,\infty}(R + K_*(T)) \\
&\leq C(R + K_*(T)) \sum_{i,j=1}^{3} (\|\varepsilon_{ij} + \tilde{\varepsilon}_{ij}\|_{\infty,\infty}\|\varepsilon_{ij} - \tilde{\varepsilon}_{ij}\|_{\infty,\infty}),
\end{aligned}
$$

so we have to look at

$$
\begin{aligned}
\varepsilon_{ij} - \tilde{\varepsilon}_{ij} &= \partial_i w_j + \partial_j w_i - \sum_{k,l=1}^{3}[(\partial_i Y_k)(\partial_k \partial_l X_j) + (\partial_j Y_k)(\partial_k \partial_l X_i)]w_l \\
&\quad - \sum_{k,l=1}^{3}[(\partial_i Y_k)(\partial_l X_j) + (\partial_j Y_k)(\partial_l X_i)]\partial_k w_l \\
&=: -\sum_{k,l=1}^{3}\kappa_{ij}^{kl}w_l + \left(\partial_i w_j - \sum_{k,l=1}^{3}(\partial_i Y_k)(\partial_l X_j)\partial_k w_l\right) \\
&\quad + \left(\partial_j w_i - \sum_{k,l=1}^{3}(\partial_j Y_k)(\partial_l X_i)\partial_k w_l\right) \\
&= \sum_{k,l=1}^{3}\left(-\kappa_{ij}^{kl}w_l + [\delta_{ik}\delta_{lj} - (\partial_i Y_k)(\partial_l X_j)]\partial_k w_l\right. \\
&\quad \left. + [\delta_{jk}\delta_{li} - (\partial_j Y_k)(\partial_l X_i)]\partial_k w_l\right) \\
&=: \sum_{k,l=1}^{3}((ii)_a + (ii)_b + (ii)_{b'}).
\end{aligned}
$$

By Proposition 2.1,

$$
\begin{aligned}
\|\delta_{ik}\delta_{lj} - (\partial_i Y_k)(\partial_l X_j)\| &\leq C\|\partial_l X_j - \delta_{lj}\|_{\infty,\infty} + C\|\partial_i Y_k - \delta_{ik}\|_{\infty,\infty} \text{ (5.19)} \\
&\leq CT\|(\xi^* + \tilde{\xi}, \Omega^* + \tilde{\Omega})\|_{\infty,\infty} \\
&\leq CT(R + C_0),
\end{aligned}
$$

and similarly

$$
\begin{aligned}
\|\partial_k \partial_l X_i\|_{\infty,\infty} &\leq C\|\partial_k(\partial_l X_j - \delta_{lj})\|_{\infty,\infty} \quad (5.20) \\
&\leq CT\|(\xi^* + \tilde{\xi}, \Omega^* + \tilde{\Omega})\|_{\infty,\infty} \\
&\leq CT(R + C_0),
\end{aligned}
$$

for all $i, j, k, l \in \{1, 2, 3\}$. Adding the corresponding estimates on Y yields

$$\|(ii)_a\|_{\infty,\infty} + \|(ii)_b\|_{\infty,\infty} + \|(ii)_{b'}\|_{\infty,\infty}$$

$$\leq \sup_{i,j,k,l} (\|\kappa_{ij}^{kl}\|_{\infty,\infty}\|w_l\|_{\infty,\infty} + \|\delta_{jk}\delta_{li} - (\partial_j Y_k)(\partial_l X_i)\|_{\infty,\infty}\|\partial_k w_l\|_{\infty,\infty})$$

$$\leq CT(R + C_0)^2,$$

so

$$\|\varepsilon_{ij} - \tilde{\varepsilon}_{ij}\|_{\infty,\infty} \leq CT(R + C_0)^2 \tag{5.21}$$

and therefore

$$\|(ii)\|_{p,p} \leq CT(K_*(T) + C_0)(R + C_0)^2. \tag{5.22}$$

In order to do (iii) in a similar way, we must first calculate

$$\partial_m \tilde{\varepsilon}_{kl}$$

$$= \frac{1}{2}\Bigg(\sum_{n,o=1}^{3} \big[(\partial_m\partial_k Y_n)(\partial_n\partial_o X_l) + (\partial_k Y_n)(\partial_m\partial_n\partial_o X_l) $$

$$+ (\partial_m\partial_l Y_n)(\partial_n\partial_o X_k) + (\partial_l Y_n)(\partial_m\partial_n\partial_o X_k) \big] w_o$$

$$+ \sum_{n,o=1}^{3} \big[(\partial_k Y_n)(\partial_n\partial_o X_l) + (\partial_l Y_n)(\partial_n\partial_o X_k) \big] \partial_m w_o$$

$$+ \sum_{n,o=1}^{3} \big[(\partial_m\partial_k Y_n)(\partial_o X_l) + (\partial_k Y_n)(\partial_m\partial_o X_l) $$

$$+ (\partial_m\partial_l Y_n)(\partial_o X_k) + (\partial_l Y_n)(\partial_m\partial_o X_k) \big] \partial_n w_o$$

$$+ \sum_{n,o=1}^{3} \big[(\partial_k Y_n)(\partial_o X_l) + (\partial_l Y_n)(\partial_o X_k) \big] \partial_m\partial_n w_o \Bigg)$$

$$=: \sum_{n,o=1}^{3} (d^{klmno} w_o + c^{klno}\partial_m w_o + \tilde{c}^{klmno}\partial_n w_o + b^{klno}\partial_m\partial_n w_o). \tag{5.23}$$

Thus, we get

$$(iii) = \Bigg(\sum_{j,k,l=1}^{3} \alpha_{ij}^{kl}\partial_j\partial_k w_l - \sum_{j,k,l,m,n,o=1}^{3} a_{ij}^{klm} b^{klno}\partial_m\partial_n w_o \Bigg)$$

$$- \Bigg(\sum_{j,k,l,m,n,o=1}^{3} a_{ij}^{klm}(c^{klno}\partial_m w_o + \tilde{c}^{klmno}\partial_n w_o) \Bigg)$$

$$- \Bigg(\sum_{j,k,l,m,n,o=1}^{3} a_{ij}^{klm} d^{klmno} w_o \Bigg)$$

$$=: (iii)_b + (iii)_c + (iii)_d.$$

It is easy to see good estimates for $(iii)_c$ and $(iii)_d$, because (5.20) can be applied to $c^{klno}, \tilde{c}^{klmno}, d^{klmno}$ and by definition,

$$\|a_{ij}^{klm}\|_{\infty,\infty} \leq C\|\mu'(|\tilde{\mathcal{E}}|_2^2)\|_{\infty,\infty}\|\nabla w\|_{\infty,\infty}^2 \leq Cm_\mu(C_0 + R)^2,$$

so that

$$\|(iii)_c\|_{\infty,\infty} + \|(iii)_d\|_{\infty,\infty} \leq CT(C_0 + R)^4. \tag{5.24}$$

We rewrite the remaining part $(iii)_b$ by using the symmetry $\alpha_{ij}^{kl} = \alpha_{ij}^{lk}$,

$$
\begin{aligned}
(iii)_b &= \frac{1}{2}\sum_{j,k,l=1}^{3} \alpha_{ij}^{kl}\partial_j\partial_k w_l - \sum_{j,k,l,m,n,o=1}^{3} a_{ij}^{klm}(\partial_k Y_n)(\partial_o X_l)\partial_m\partial_n w_o \\
&\quad + \frac{1}{2}\sum_{j,k,l=1}^{3} \alpha_{ij}^{kl}\partial_j\partial_k w_l - \sum_{j,k,l,m,n,o=1}^{3} a_{ij}^{klm}(\partial_l Y_n)(\partial_o X_k)\partial_m\partial_n w_o \\
&= \sum_{j,k,l,m,n,o=1}^{3} [\frac{1}{2}\alpha_{ij}^{kl}\delta_{jm}\delta_{nk}\delta_{ol} - a_{ij}^{klm}(\partial_k Y_n)(\partial_o X_l)]\partial_m\partial_n w_o \\
&\quad + \sum_{j,k,l,m,n,o=1}^{3} [\frac{1}{2}\alpha_{ij}^{kl}\delta_{jm}\delta_{nl}\delta_{ok} - a_{ij}^{klm}(\partial_k Y_n)(\partial_o X_l)]\partial_m\partial_n w_o \\
&= \sum_{j,k,l,m,n,o=1}^{3} \Big[2\mu'(|\mathcal{E}|_2^2)\varepsilon_{ij}\varepsilon_{kl}\delta_{jm}\delta_{nk}\delta_{ol} \\
&\quad -2\mu'(|\tilde{\mathcal{E}}|_2^2)\tilde{\varepsilon}_{ij}\tilde{\varepsilon}_{kl}(\partial_j Y_m)(\partial_k Y_n)(\partial_o X_l)\Big]\partial_m\partial_n w_o \\
&\quad + \sum_{j,k,l,m,n,o=1}^{3} \Big[2\mu'(|\mathcal{E}|_2^2)\varepsilon_{ij}\varepsilon_{kl}\delta_{jm}\delta_{nk}\delta_{ol} \\
&\quad -2\mu'(|\tilde{\mathcal{E}}|_2^2)\tilde{\varepsilon}_{ij}\tilde{\varepsilon}_{kl}(\partial_j Y_m)(\partial_k Y_n)(\partial_o X_l)\Big]\partial_m\partial_n w_o.
\end{aligned}
$$

It is now clear from the structure of this term that by adding zeros as in the terms above and using the estimates (5.21) and (5.19) it follows that

$$\|(iii)_b\|_{\infty,\infty} \leq CT(C_0 + R). \tag{5.25}$$

In conclusion, from (5.18), (5.22), (5.24) and (5.25) we obtain

$$\|\mathcal{Q}_{III}\|_{p,p} \leq CT(C_0 + R).$$

Putting together the estimates on $\mathcal{Q}_I$, $\mathcal{Q}_{II}$ and $\mathcal{Q}_{III}$ yields

$$\|\mathcal{Q}(u^*, \tilde{u})\|_{p,p} \leq C(R^2 + K_*(T) + T). \tag{5.26}$$

This is the main estimate we need in addition to the results from Chapter 4 to prove the first part of the following lemma.

Lemma 5.4. *For sufficiently small $T > 0$ and $R > 0$, the image $\mathrm{Im}(\psi_R^T)$ of ψ_R^T is contained in $\mathcal{K}_R^T$ and ψ_R^T is contractive.*

Proof. In Lemmas 4.3 and 4.4, estimates on F_0, F_1, F_2 were shown. We use that these functions are parts of G_0, G_1 and G_2. By definition and by (5.26),

$$\begin{aligned}
\|G_0(\tilde{u}, \tilde{p}, \tilde{\xi}, \tilde{\Omega})\|_{p,p} &\leq \|F_0(\tilde{u}, \tilde{p}, \tilde{\xi}, \tilde{\Omega})\|_{p,p} + K_*(T) + \|\mathcal{Q}(\tilde{u}, u^*)\|_{p,p} \\
&\leq C(T^{1/2} + T^{1/p} + R^2 + K_*(T)).
\end{aligned}$$

In the following, let Q be the matrix which corresponds to the velocity $\Omega = \tilde{\Omega} + \Omega^*$, cf. step (1) in Section 2.3. By definition,

$$\begin{aligned}
\|G_1(\tilde{u}, \tilde{p}, \tilde{\xi}, \tilde{\Omega})\|_p &\leq \|F_1(\tilde{u}, \tilde{p}, \tilde{\xi}, \tilde{\Omega})\|_p + \left\| \int_\Gamma (\mathcal{T} - \mathcal{T}^\mu)(\tilde{u}, \tilde{p}) N \, d\sigma \right\|_p \\
&\quad + \left\| \int_\Gamma (\mathbf{T} - \mathcal{T}^\mu)(u^*, p^*) N \, d\sigma \right\|_p \\
&\leq CT^{1/p} + C\|\mathcal{J}(Q^T(\mu_0 - \mu(|\mathcal{E}^{(\tilde{u})}|_2^2))\mathcal{E}^{(Q\tilde{u})}Q)\|_p \\
&\quad + \|\mathcal{J}(\mu_0\mathcal{E}^{(u^*)} - Q^T\mu(|\mathcal{E}^{(u^*)}|_2^2)\mathcal{E}^{(Qu^*)}Q)\|_p \\
&\leq CT^{1/p} + C\|\mathcal{E}^{(Q(u^*))}\|_{L^p(J_T;C(\mathcal{D}))} + \|\mathcal{E}^{(Q(\tilde{u}))}\|_{L^p(J_T;C(\mathcal{D}))} \\
&\leq CT^{1/p}
\end{aligned}$$

and similarly,

$$\begin{aligned}
\|G_2(\tilde{u}, \tilde{p}, \tilde{\Omega})\|_p &\leq \|F_2(\tilde{p}, \tilde{\xi}, \tilde{\Omega})\|_p + \|\mathcal{J}((\mathcal{T} - \mathcal{T}^\mu)(\tilde{u}, \tilde{p}))\|_p \\
&\quad + \|\mathcal{J}((\mathbf{T} - \mathcal{T}^\mu)(u^*, p^*))\|_p \\
&\leq CT^{1/p}.
\end{aligned}$$

Thus by Theorem 5.2,

$$\begin{aligned}
\|\psi_R^T(\tilde{u}, \tilde{p}, \tilde{\xi}, \tilde{\Omega})\|_{\mathcal{K}_R^T} &\leq C(\|G_0(\tilde{u}, \tilde{p}, \tilde{\xi}, \tilde{\Omega})\|_{p,p} + \|(G_1(\tilde{u}, \tilde{p}, \tilde{\xi}, \tilde{\Omega}), G_2(\tilde{u}, \tilde{p}, \tilde{\Omega}))\|_p) \\
&\leq R \qquad \text{for sufficiently small } T, R.
\end{aligned}$$

Let now $(\tilde{u}_1, \tilde{p}_1, \tilde{\xi}_1, \tilde{\Omega}_1), (\tilde{u}_2, \tilde{p}_2, \tilde{\xi}_2, \tilde{\Omega}_2) \in \mathcal{K}_R^T$. In the following, we again use the convention from Section 2.3 and put an index 1 or 2 on a function or an operator to indicate that it is constructed using either the change of coordinates X_1 corresponding to $\tilde{\xi}_1, \tilde{\Omega}_1$ or using X_2 corresponding to $\tilde{\xi}_2, \tilde{\Omega}_2$, respectively. Note that this does not apply to the functions F_0, F_1, F_2, G_0, G_1 and G_2. We add $\pm A(u^* + \tilde{u}_1)(\tilde{u}_1 - \tilde{u}_2)$ and $\pm \mathcal{A}_1(u^* + \tilde{u}_1)\tilde{u}_2$ to $\mathcal{Q}_1(u^*, \tilde{u}_1) -$

$\mathcal{Q}_2(u^*, \tilde{u}_2)$ to define

$$\begin{aligned}
\|\mathcal{Q}_1(u^*, \tilde{u}_1) - \mathcal{Q}_2(u^*, \tilde{u}_2)\|_{p,p} &\leq \|[A_* - A(u^* + \tilde{u}_1)](\tilde{u}_1 - \tilde{u}_2)\|_{p,p} \\
&\quad + \|[A(u^* + \tilde{u}_1) - \mathcal{A}_1(u^* + \tilde{u}_1)](\tilde{u}_1 - \tilde{u}_2)\|_{p,p} \\
&\quad + \|[\mathcal{A}_2(u^* + \tilde{u}_2) - \mathcal{A}_1(u^* + \tilde{u}_1)](u^* + \tilde{u}_2)\|_{p,p} \\
&=: q_I + q_{II} + q_{III}.
\end{aligned}$$

From the estimates for $\mathcal{Q}_I$ and $\mathcal{Q}_{III}$ we can see that

$$\|q_I\|_{p,p} + \|q_{II}\|_{p,p} \leq C(R+T)\|\tilde{u}_1 - \tilde{u}_2\|_{X_{p,p}^T}. \tag{5.27}$$

Let now $w_1 := \tilde{u}_1 + u^*$ and $w_2 := \tilde{u}_2 + u^*$ to make the estimates shorter. We split the remaining term q_{III} as follows,

$$\begin{aligned}
\Big([\mathcal{A}_2(w_2) - \mathcal{A}_1(w_1)](w_2)\Big)_i &= \Big([\mu(|\tilde{\mathcal{E}}_2^{(w_2)}|_2^2) - \mu(|\tilde{\mathcal{E}}_2^{(w_1)}|_2^2)](\mathcal{L}_2 w_2)\Big)_i \\
&\quad + \Big(\mu(|\tilde{\mathcal{E}}_1^{(w_1)}|_2^2)(\mathcal{L}_2 - \mathcal{L}_1)(w_2)\Big)_i \\
&\quad + \sum_{j,k,l,m=1}^{3} ((a_2)_{ij}^{klm}(w_2) - (a_1)_{ij}^{klm}(w_1))\partial_m(\tilde{\varepsilon}_2)_{kl}^{(w_2)} \\
&\quad + \sum_{j,k,l,m=1}^{3} (a_1)_{ij}^{klm}(w_1)\partial_m((\tilde{\varepsilon}_2)_{kl}^{(w_2)} - (\tilde{\varepsilon}_1)_{kl}^{(w_1)}) \\
&=: A_I + A_{II} + A_{III} + A_{IV}.
\end{aligned}$$

From the definition in (5.6) we obtain

$$\begin{aligned}
&\|(\tilde{\varepsilon}_1)_{ij}^{(w_1)} - (\tilde{\varepsilon}_2)_{ij}^{(w_2)}\|_{\infty,\infty} \\
&\leq \sup_{k,l}\Big[\|(\tilde{e}_1)_{kl}^{ij} - (\tilde{e}_2)_{kl}^{ij}\|_{\infty,\infty}\|w_1\|_{\infty,\infty} + \|(\tilde{e}_2)_{kl}^{ij}\|_{\infty,\infty}\|\tilde{u}_1 - \tilde{u}_2\|_{\infty,\infty} \\
&\quad + \|(\tilde{d}_1)_{kl}^{ij} - (\tilde{d}_2)_{kl}^{ij}\|_{\infty,\infty}\|\nabla w_1\|_{\infty,\infty} + \|(\tilde{d}_2)_{kl}^{ij}\|_{\infty,\infty}\|\nabla(\tilde{u}_1 - \tilde{u}_2)\|_{\infty,\infty}\Big]
\end{aligned}$$

and Proposition 4.2 shows that

$$\begin{aligned}
\|(\tilde{e}_1)_{kl}^{ij} - (\tilde{e}_2)_{kl}^{ij}\|_{\infty,\infty} &\leq \|\partial_i(Y_1)_k\|_{\infty,\infty}\|\partial_k\partial_l(X_1 - X_2)_j\|_{\infty,\infty} \\
&\quad + \|\partial_i((Y_1)_k(\cdot, X_1) - (Y_2)_k(\cdot, X_2))\|_{\infty,\infty}\|\partial_k\partial_l(X_2)_j\|_{\infty,\infty} \\
&\leq CT\|(\tilde{\xi}_1 - \tilde{\xi}_2, \tilde{\Omega}_1 - \tilde{\Omega}_2)\|_{W^{1,p}(J_T)}
\end{aligned}$$

and

$$\|(\tilde{d}_1)_{kl}^{ij} - (\tilde{d}_2)_{kl}^{ij}\|_{\infty,\infty} \leq CT\|(\tilde{\xi}_1 - \tilde{\xi}_2, \tilde{\Omega}_1 - \tilde{\Omega}_2)\|_{W^{1,p}(J_T)}.$$

Hence,

$$\|(\tilde{\varepsilon}_1)_{ij}^{(w_1)} - (\tilde{\varepsilon}_2)_{ij}^{(w_2)}\|_{\infty,\infty} \le C(\|(\tilde{\xi}_1 - \tilde{\xi}_2, \tilde{\Omega}_1 - \tilde{\Omega}_2)\|_{W^{1,p}(J_T)} + \|\tilde{u}_1 - \tilde{u}_2\|_{X_{p,p}^T})$$

and it follows that

$$\begin{aligned}
&\|\mu'(|\tilde{\mathcal{E}}_2^{(w_2)}|_2^2) - \mu'(|\tilde{\mathcal{E}}_1^{(w_1)}|_2^2)\|_{\infty,\infty} \\
\le\ & Cm_\mu \sup_{i,j} \|(\tilde{\varepsilon}_1)_{ij}^{(w_1)} + (\tilde{\varepsilon}_2)_{ij}^{(w_2)}\|_{\infty,\infty} \|(\tilde{\varepsilon}_1)_{ij}^{(w_1)} - (\tilde{\varepsilon}_2)_{ij}^{(w_2)}\|_{\infty,\infty} \\
\le\ & C(\|(\tilde{\xi}_1 - \tilde{\xi}_2, \tilde{\Omega}_1 - \tilde{\Omega}_2)\|_{W^{1,p}(J_T)} + \|\tilde{u}_1 - \tilde{u}_2\|_{X_{p,p}^T})
\end{aligned}$$

and therefore

$$\begin{aligned}
&\|(a_1)_{ij}^{klm}(w_1) - (a_2)_{ij}^{klm}(w_2)\|_{\infty,\infty} \\
\le\ & \|2[\mu'(|\tilde{\mathcal{E}}_2^{(w_2)}|_2^2) - \mu'(|\tilde{\mathcal{E}}_1^{(w_1)}|_2^2)]\partial_j(Y_2)_m(\tilde{\varepsilon}_2)_{ij}^{(w_2)}(\tilde{\varepsilon}_2)_{kl}^{(w_2)}\|_{\infty,\infty} \\
&+ \|2\mu'(|\tilde{\mathcal{E}}_1^{(w_1)}|_2^2)[\partial_j(Y_2)_m - \partial_j(Y_1)_m](\tilde{\varepsilon}_2)_{ij}^{(w_2)}(\tilde{\varepsilon}_2)_{kl}^{(w_2)}\|_{\infty,\infty} \\
&+ \|2\mu'(|\tilde{\mathcal{E}}_1^{(w_1)}|_2^2)\partial_j(Y_1)_m[(\tilde{\varepsilon}_2)_{ij}^{(w_2)} - (\tilde{\varepsilon}_1)_{ij}^{(w_1)}](\tilde{\varepsilon}_2)_{kl}^{(w_2)}\|_{\infty,\infty} \\
&+ \|2\mu'(|\tilde{\mathcal{E}}_1^{(w_1)}|_2^2)\partial_j(Y_1)_m(\tilde{\varepsilon}_2)_{ij}^{(w_2)}[(\tilde{\varepsilon}_2)_{kl}^{(w_2)} - (\tilde{\varepsilon}_1)_{kl}^{(w_1)}]\|_{\infty,\infty} \\
\le\ & C(\|(\tilde{\xi}_1 - \tilde{\xi}_2, \tilde{\Omega}_1 - \tilde{\Omega}_2)\|_{W^{1,p}(J_T)} + \|\tilde{u}_1 - \tilde{u}_2\|_{X_{p,p}^T}).
\end{aligned}$$

Secondly, by the definition of $\mathcal{L}$ in (2.18) and by (5.23) we get

$$\|\mathcal{L}_2 w_2\|_{p,p} + \sup_{k,l,m} \|\partial_m(\tilde{\varepsilon}_2)_{kl}^{(w_2)}\|_{p,p} \le C\|w_2\|_{X_{p,p}^T} \le C(R + K_*(T)).$$

As a direct consequence, we obtain

$$\|A_I\|_{p,p} + \|A_{III}\|_{p,p} \le C(R + K_*(T))(\|(\tilde{\xi}_1 - \tilde{\xi}_2, \tilde{\Omega}_1 - \tilde{\Omega}_2)\|_{W^{1,p}(J_T)} + \|\tilde{u}_1 - \tilde{u}_2\|_{X_{p,p}^T}).$$

Moreover, the estimate

$$\|A_{II}\|_{p,p} \le CT\|(\tilde{\xi}_1 - \tilde{\xi}_2, \tilde{\Omega}_1 - \tilde{\Omega}_2)\|_{W^{1,p}(J_T)}$$

follows directly from (4.5).

In a similar way, we do A_{IV}. Clearly, the coefficients in (5.23) satisfy

$$\begin{aligned}
\sup_{k,l,n,o} \|(b_1 - b_2)^{klno}\|_{\infty,\infty} \le\ & \sup_{k,l,n,o}\Big[\|\partial_k(Y_1 - Y_2)_n\|_{\infty,\infty}\|\partial_o(X_1)_l\|_{\infty,\infty} \\
&+ \|\partial_k(Y_2)_n\|_{\infty,\infty}\|\partial_o(X_1 - X_2)_l\|_{\infty,\infty}\Big] \\
\le\ & CT\|(\tilde{\xi}_1 - \tilde{\xi}_2, \tilde{\Omega}_1 - \tilde{\Omega}_2)\|_\infty
\end{aligned}$$

and similarly

$$\sup_{k,l,m,n,o} \left[\|(c_1 - c_2)^{klno}\|_{\infty,\infty} + \|(\tilde{c}_1 - \tilde{c}_2)^{klmno}\|_{\infty,\infty} + \|(d_1 - d_2)^{klmno}\|_{\infty,\infty} \right]$$
$$\leq CT\|(\tilde{\xi}_1 - \tilde{\xi}_2, \tilde{\Omega}_1 - \tilde{\Omega}_2)\|_\infty$$

by Proposition 2.1. These estimates imply

$$\|A_{IV}\|_{p,p} \leq C \sup_{k,l,m} \|\partial_m((\tilde{\varepsilon}_2 - \tilde{\varepsilon}_1)_{kl}^{(w_2)})\|_{p,p}$$
$$\leq C \sup_{k,l,m,n,o} \Big(\|(b_1 - b_2)^{klno}\|_{\infty,\infty} + \|(c_1 - c_2)^{klno}\|_{\infty,\infty}$$
$$+ \|(\tilde{c}_1 - \tilde{c}_2)^{klmno}\|_{\infty,\infty} + \|(d_1 - d_2)^{klmno}\|_{\infty,\infty} \Big) \|w_2\|_{X_{p,p}^T}$$
$$\leq CT(R + K_*(T))\|(\tilde{\xi}_1 - \tilde{\xi}_2, \tilde{\Omega}_1 - \tilde{\Omega}_2)\|_{W^{1,p}(J_T)}.$$

In conclusion,

$$\|q_{III}\|_{p,p}$$
$$\leq \|A_I\|_{p,p} + \|A_{II}\|_{p,p} + \|A_{III}\|_{p,p} + \|A_{IV}\|_{p,p}$$
$$\leq C(K_*(T) + R + T)(\|\tilde{u}_1 - \tilde{u}_2\|_{X_{p,p}^T} + \|(\tilde{\xi}_1 - \tilde{\xi}_2, \tilde{\Omega}_1 - \tilde{\Omega}_2)\|_{W^{1,p}(J_T)})$$

and therefore

$$\|\mathcal{Q}_1(u^*, \tilde{u}_1) - \mathcal{Q}_2(u^*, \tilde{u}_2)\|_{p,p}$$
$$\leq \|q_I\|_{p,p} + \|q_{II}\|_{p,p} + \|q_{III}\|_{p,p}$$
$$\leq C(K_*(T) + R + T)(\|\tilde{u}_1 - \tilde{u}_2\|_{X_{p,p}^T} + \|(\tilde{\xi}_1 - \tilde{\xi}_2, \tilde{\Omega}_1 - \tilde{\Omega}_2)\|_{\infty,\infty}),$$

where we additionally used (5.27).

From Lemma 4.2, the estimates in the proof of Lemma 4.4 and the estimate above we get

$$\|G_0(\tilde{u}_1, \tilde{p}_1, \tilde{\xi}_1, \tilde{\Omega}_1) - G_0(\tilde{u}_2, \tilde{p}_2, \tilde{\xi}_2, \tilde{\Omega}_2)\|_{p,p}$$
$$\leq L_{R,T}(\|\tilde{u}_1 - \tilde{u}_2\|_{X_{p,p}^T} + \|\tilde{p}_1 - \tilde{p}_2\|_{Y_{p,p}^T} + \|(\tilde{\xi}_1 - \tilde{\xi}_2, \tilde{\Omega}_1 - \tilde{\Omega}_2)\|_{\infty,\infty}),$$

where we choose $L_{R,T} = C(K_*(T) + R + T + T^{1/p})$.

Next we consider the right hand sides of the body equations, G_1 and G_2. In particular, it remains to show that

$$\mathcal{J}^\mu := \|\mathcal{J}((\mathcal{T}_1 - \mathcal{T}_1^\mu)(\tilde{u}_1, \tilde{p}_1) - (\mathcal{T}_2 - \mathcal{T}_2^\mu)(\tilde{u}_2, \tilde{p}_2))\|_{p,p}$$
$$\leq CT\big(\|\tilde{u}_1 - \tilde{u}_2\|_{X_{p,q}^T} + \|\tilde{\Omega}_1 - \tilde{\Omega}_2\|_{W^{1,p}(J_T)}\big).$$

Note that the pressure terms $\tilde{p}_1$ and $\tilde{p}_2$ disappear since $\mathcal{T}$ and $\mathcal{T}^\mu$ only differ in the viscosity μ. If we write

$$\mathcal{J}\big[(\mathcal{T}_1 - \mathcal{T}_2)(\tilde{u}_1, \tilde{p}_1) + (\mathbf{T} - \mathcal{T}_2)(\tilde{u}_1 - \tilde{u}_2, \tilde{p}_1 - \tilde{p}_2) \\ -(\mathcal{T}_1^\mu - \mathcal{T}_2^\mu)(\tilde{u}_1, \tilde{p}_1) - (\mathbf{T} - \mathcal{T}_2^\mu)(\tilde{u}_1 - \tilde{u}_2, \tilde{p}_1 - \tilde{p}_2)\big]$$

instead of the term in the first line, it is clear that we can use exactly the same proof as in Lemma 4.4 if we replace $\mathcal{T}$ by $\mathcal{T}^\mu$ and use the estimates on the differences of the viscosity terms in case of two different transforms and given velocities obtained above. In addition, it follows that

$$\mathcal{J}_*^\mu \;\; := \;\; \|\mathcal{J}((\mathcal{T}_1^\mu - \mathcal{T}_2^\mu)(u^*, p^* - g \cdot y))\|_p \leq CT\|\tilde{\Omega}_1 - \tilde{\Omega}_2\|_{W^{1,p}(J_T)}$$

as in Lemma 4.2. It follows from the definitions of G_1 and G_2 that

$$\|G_1(\tilde{u}_1, \tilde{p}_1, \tilde{\xi}_1, \tilde{\Omega}_1) - G_1(\tilde{u}_2, \tilde{p}_2, \tilde{\xi}_2, \tilde{\Omega}_2)\|_p$$
$$\leq \;\; \|F_1(\tilde{u}_1, \tilde{p}_1, \tilde{\xi}_1, \tilde{\Omega}_1) - F_1(\tilde{u}_2, \tilde{p}_2, \tilde{\xi}_2, \tilde{\Omega}_2)\|_p + \mathcal{J}^\mu + \mathcal{J}_*^\mu$$
$$\leq \;\; CT\|(\tilde{u}_1 - \tilde{u}_2, \tilde{p}_1 - \tilde{p}_2, \tilde{\xi}_1 - \tilde{\xi}_2, \tilde{\Omega}_1 - \tilde{\Omega}_2)\|_{\mathcal{K}_R^T}$$

and

$$\|G_2(\tilde{u}_1, \tilde{p}_1, \tilde{\Omega}_1) - G_2(\tilde{u}_2, \tilde{p}_2, \tilde{\Omega}_2)\|_p$$
$$\leq \;\; \|F_2(\tilde{u}_1, \tilde{p}_1, \tilde{\Omega}_1) - F_2(\tilde{u}_2, \tilde{p}_2, \tilde{\Omega}_2)\|_p + \mathcal{J}^\mu + \mathcal{J}_*^\mu$$
$$\leq \;\; CT\|(\tilde{u}_1 - \tilde{u}_2, \tilde{p}_1 - \tilde{p}_2, \tilde{\xi}_1 - \tilde{\xi}_2, \tilde{\Omega}_1 - \tilde{\Omega}_2)\|_{\mathcal{K}_R^T}.$$

This implies

$$\|\psi_R^T(\tilde{u}_1, \tilde{p}_1, \tilde{\xi}_1, \tilde{\Omega}_1) - \psi_R^T(\tilde{u}_2, \tilde{p}_2, \tilde{\xi}_2, \tilde{\Omega}_2)\|_{\mathcal{K}_R^T}$$
$$\leq \;\; C\big(\|G_0(\tilde{u}_1, \tilde{p}_1, \tilde{\xi}_1, \tilde{\Omega}_1) - G_0(\tilde{u}_2, \tilde{p}_2, \tilde{\xi}_2, \tilde{\Omega}_2)\|_{p,p}$$
$$+ \;\; \|G_1(\tilde{u}_1, \tilde{p}_1, \tilde{\xi}_1, \tilde{\Omega}_1) - G_1(\tilde{u}_2, \tilde{p}_2, \tilde{\xi}_2, \tilde{\Omega}_2)\|_p$$
$$+ \;\; \|G_2(\tilde{u}_1, \tilde{p}_1, \tilde{\Omega}_1) - G_2(\tilde{u}_2, \tilde{p}_2, \tilde{\Omega}_2)\|_p\big)$$
$$\leq \;\; C(R + K_*(T) + T^{1/p} + T^{1/2})\|(\tilde{u}_1 - \tilde{u}_2, \tilde{p}_1 - \tilde{p}_2, \tilde{\xi}_1 - \tilde{\xi}_2, \tilde{\Omega}_1 - \tilde{\Omega}_2)\|_{\mathcal{K}_R^T}.$$

And thus for sufficiently small R, T, the map ψ_R^T becomes a contraction. $\square$

From the Banach fixed point theorem applied to ψ_R^T, it follows that there is a unique strong solution $(\hat{u}, \hat{p}, \hat{\xi}, \hat{\Omega})$ to problem (5.10). The solution to the original problem (5.4) can be found by adding the reference solution $(u^*, p^* + g \cdot y, \xi^*, \Omega^*)$ and performing the corresponding backward coordinate transform, exactly as in the proof of Theorem 4.6.

Chapter 6

Similar Results

In this chapter we give a formulation of the model problems in a bounded container and in two space dimensions and show that in these cases, similar results can be obtained.

Consider the situation of a fixed bounded container $\mathcal{O}$ which holds both the rigid body in a bounded domain $\mathcal{B}(t)$ and the liquid. On the one hand, it is a positive feature of the non-linear change of coordinates X we use in Chapter 2 that it can be modified to leave $\mathcal{O}$ fixed and at the same time transform $\mathcal{B}(t)$ back to $\mathcal{B}$ due to the fact that it can be "cut off" to resemble the identity in the complement of arbitrarily small balls wich contain $\mathcal{B}$. On the other hand, this setup generates the problem of possible contact of the "inner" and "outer" boundaries which is difficult to understand. For example, if we want to find strong solutions and maintain no-slip boundary conditions, it could happen that the body and the wall can never touch, cf. the result in [HT09] on the special case of a sphere moving over a plain and cf. [FHN08] for the case of generalized Newtonian, shear-thickening fluids. For the weak formulation of the Newtonian problem, it can be shown that global solutions exist regardless of contact, if several obstacles move in a bounded domain of sufficient regularity, cf. [Fei03]. These considerations are out of the reach of our method. The change of coordinates we use becomes unfeasible if two boundaries get too close, so the contact problem has to be avoided by limiting the solution to sufficiently small time intervals. A similar restriction is used by Takahashi in [Tak03]. In the L^2-setting, the energy estimates from [DE99] can be used to estimate the displacement of the rigid body in terms of the initial data. Sufficiently small data thus guarantee the existence of a global solution to the problem and that the body never reaches the boundary. Note that this result does not necessarily follow if we consider the bulk force $\mathbf{g}$

acting on body and fluid, but it requires the exterior force to vanish after some time, cf. [Tak03, Corollary 9.2].

It is clearer to see how to extend the main results from Chapters 4 and 5 to the model in two space dimensions. The abstract preliminary maximal regularity results on the fluid part of the problem are known in general dimensions $n \geq 2$, so they can still be used. The balance equations for the body become simpler as in two dimensions, "fewer" rigid motions of the body are possible.

Remark 6.1. A generalization of the main results which can be done easily is to allow for external forces and torques different from $\mathbf{g}$, which do not change the solvability of the reference problems (4.1) and (5.8). External forces on the fluid which can be included on the right hand side of the first line in (2.1) have to satisfy $f_{\text{ext}} \in L^p(J_T; L^q(\mathbb{R}^3))$ and the functions $\mathbf{F}, \mathbf{M}$ in the balance equations for the rigid body can be chosen from $L^p(J_T)$, if $1 < p, q < \infty$ satisfy the assumptions of the corresponding theorem.

6.1   Bounded Fluid Domains

Let $\mathcal{O} \subset \mathbb{R}^3$ be a bounded domain of class $C^{2,1}$. We consider $\mathcal{O}$ as a container for both the body and the fluid. The notation and setup of this model is illustrated in the figure below.

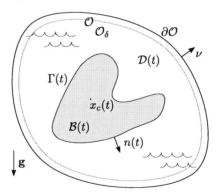

Again, $\mathcal{B}(t) \subset \mathcal{O}$ is the body's domain and the liquid fills $\mathcal{D}(t) = \mathcal{O} \backslash \overline{\mathcal{B}(t)}$. The domains start out from $\mathcal{D} := \mathcal{D}(0)$ and $\mathcal{B} := \mathcal{B}(0)$ and then move as functions of the body's velocity. The interface between body and fluid is denoted by $\Gamma(t)$, so that $\partial \mathcal{D}(t) = \Gamma(t) \cup \partial \mathcal{O}$. The function $n(t)$ denotes the outer normal of $\mathcal{B}(t)$ at $\Gamma(t)$ and ν is the outer normal of $\mathcal{O}$.

It is an advantage of our change of coordinates given by X and Y from Chapter 2 that it can be defined to leave the boundary $\partial\mathcal{O}$ fixed and still rotate and shift the body back to its original position at every time $0 \leq t < T$. This is not possible under the linear transforms given by X_0 and Y_0, as $\mathcal{O}$ is not assumed to be invariant under rotation and translation.

Since we cannot deal with possible contact of the body and the boundary of $\mathcal{O}$, we assume that the body starts from a position with some distance from $\partial\mathcal{O}$, i.e. $\operatorname{dist}(\Gamma(0), \partial\mathcal{O}) = d_0 > 0$. Since it moves with a continuous velocity, we can restrict the solution to an interval of time which guarantees that a small distance δ, $d_0 > \delta > 0$, remains between body and wall. In the following, let $\mathcal{O}_\delta = \{x \in \mathcal{O} : \operatorname{dist}(x, \partial\mathcal{O}) > \delta\}$.

Otherwise, we use the same notation and model as in Section 2.1. We assume that $x_c(0) = 0$ for the position of the center of gravity of the body at time 0, so that $0 \in \mathcal{B}$ and $0 \in \mathcal{O}$. Moreover, let $r_\mathcal{B} := \inf_{r>0}\{\mathcal{B} \subset B_r\}$ the extension of $\mathcal{B}$ from its center of gravity. Including Dirichlet boundary conditions on $\partial\mathcal{O}$, the initial-boundary value problem reads

$$\begin{cases} v_t - \operatorname{div}\mathbf{T}(v,q) + (v \cdot \nabla)v &= \mathbf{g} & \text{in } J_T \times \mathcal{D}(t), \\ \operatorname{div}v &= 0 & \text{in } J_T \times \mathcal{D}(t), \\ v &= v_\mathcal{B} & \text{on } J_T \times \Gamma(t), \\ v &= 0 & \text{on } J_T \times \partial\mathcal{O}, \\ v(0) &= v_0 & \text{in } \mathcal{D}, \\ \mathrm{m}\eta' + \int_{\Gamma(t)}\mathbf{T}(v,q)n\,\mathrm{d}\sigma, &= \mathrm{m}g & \text{in } J_T, \\ (J\omega)' + \int_{\Gamma(t)}(x - x_c) \times \mathbf{T}(v,q)n\,\mathrm{d}\sigma &= 0 & \text{in } J_T, \\ \eta(0) &= \eta_0, \\ \omega(0) &= \omega_0, \end{cases} \qquad (6.1)$$

where $v_\mathcal{B}(t,x) = \eta(t) + \omega(t) \times (x - x_c(t))$ for all $(t,x) \in J_T \times \mathcal{B}(t)$. Concerning this system of equations, the following result holds true.

Theorem 6.2. *Let $\mathcal{O}$, $\mathcal{D}$, $\mathcal{B}$ be bounded domains of class $C^{2,1}$ in $\mathbb{R}^3$ as described above. Assume that $0 < \delta < d_0$, $\frac{3}{2q} + \frac{1}{p} \leq \frac{3}{2}$, $\frac{1}{2q} + \frac{1}{p} \neq 1$, $\eta_0, \omega_0 \in \mathbb{R}^3$ and that $v_0 \in B_{q,p}^{2-2/p}(\mathcal{D})$ satisfies the compatibility conditions $\operatorname{div}v_0 = 0$ and (3.3) on Γ and $v_0|_{\partial\mathcal{O}} = 0$, if $\frac{1}{2q} + \frac{1}{p} < 1$ and $v_0|_{\partial\mathcal{O}} \cdot \nu = 0$, if $\frac{1}{2q} + \frac{1}{p} > 1$ on $\partial\mathcal{O}$.*

Then there exists a maximal interval J_{T_δ}, $T_\delta > 0$, such that problem (6.1) *admits a unique strong solution*

$$v \quad \in \quad L^p(J_{T_\delta}; W^{2,q}(\mathcal{D}(\cdot))) \cap W^{1,p}(J_{T_\delta}; L^q(\mathcal{D}(\cdot))),$$

$$q \quad = \quad q_0 + \mathbf{g} \cdot Y, \quad q_0 \in L^p(J_{T_\delta}; \widehat{W}^{1,q}(\mathcal{D}(\cdot))), Y \in C^1(J_{T_\delta}; C^\infty(\mathcal{D}(\cdot))),$$

$$\eta \quad \in \quad W^{1,p}(J_{T_\delta}),$$

$$\omega \quad \in \quad W^{1,p}(J_{T_\delta}).$$

Moreover, T_δ is characterized by the following conditions. Either

1. *$T_\delta = +\infty$ or*

2. *one of the functions $t \mapsto \|v(t)\|_{B^{2-2/p}_{q,p}(\mathcal{D}(t))}$, $t \mapsto |\eta(t)|$, $t \mapsto |\omega(t)|$ is unbounded on J_{T_δ} or*

3. *the body enters $\mathcal{O}_\delta$, i.e. the distance*

$$\mathrm{dist}(\mathcal{B}(t), \partial\mathcal{O}) \leq d_0 - |\int_0^{T_\delta} \eta(t)\,\mathrm{d}t| - r_\mathcal{B}$$

becomes smaller than δ.

The result follows with the methods used in Chapters 2-4 to prove Theorem 4.6. It would be too tedious to repeat all the details, but in the following, we go through the crucial arguments and explain shortly how the additional boundary condition fits in.

First the change of coordinates from Chapter 2 has to be modified slightly so that the cut-off function χ is guaranteed to respect the outer boundary. Let $\overline{\chi} \in C^\infty(\mathbb{R}^3; [0,1])$ be given by

$$\overline{\chi}(x) := \begin{cases} 1 & \text{if } x \in \mathcal{O}_\delta, \\ 0 & \text{if } x \in \mathcal{O} \backslash \mathcal{O}_{\frac{\delta}{2}} \end{cases}$$

and $b : [0,T] \times \mathbb{R}^3 \to \mathbb{R}^3$ by

$$b(t,x) := \overline{\chi}(x)[m(t)(x - x_c(t)) + \eta(t)] - B_{\mathcal{O}}(\nabla\overline{\chi}(\cdot)[m(t)(\cdot - x_c(t)) + \eta(t)])(x).$$

As in Section 2.1, this leads to transforms $X(t)$ and $Y(t)$ of $\mathbb{R}^3$ which act as volume-preserving diffeomorphisms on $\mathcal{O}$. The transformed system of equations in the transformed unknowns u, p, ξ, Ω on the cylindrical domain $J_T \times \mathcal{D}$ is like (2.17), except for the additional condition

$$u|_{\partial\mathcal{O}} = 0. \tag{6.2}$$

The corresponding linearized system is also the same as (3.1) and includes (6.2).

The maximal regularity results on the linear problem can be reproduced by using the same arguments as in Chapter 3. The properties of the fluid domain enter into the definition of the solution operators $\mathcal{U}$ and $\mathcal{P}$ to the Stokes problem with inhomogeneous boundary conditions on $\mathcal{D}$. Analogously to the procedure in Section 1.5, the linear problem with boundary conditions (6.2) and $u|_\Gamma = h$, where h is defined as in (3.9), can be reduced to the Stokes problem with homogeneous boundary conditions on the bounded domain $\mathcal{D}$. Proposition 1.5 also applies in this situation because it allows for general domains of class C^2 with compact boundaries. The pressure estimate in Lemma 1.10 also holds true for all $\mathcal{P}(f, 0, 0)$ chosen from $L_0^q(\mathcal{D})$, so that maximal regularity of the linear problem on a bounded domain follows analogously to Theorem 3.1.

The estimates on the contraction mapping ϕ_R^T from Chapter 4 can be preserved as the main ingredients, the embedding properties of $X_{p,q}^T$ and $_0W^{1,p}(J_T)$ from Section 1.4 do not change.

From the properties of the solution v, q, η, ω it follows that the interval J_{T_δ} is well-defined by the argument used in the proof of Theorem 4.6. It is only maximal under the restriction that we want to avoid the contact problem and therefore require the parameter δ.

Remark 6.3. In the same way, the proof goes through in the generalized Newtonian situation. The definition of the operators $\mathcal{U}_{A_*}$ and $\mathcal{P}_{A_*}$ in Subsection 1.3.2 is derived from Proposition 1.7 on maximal regularity of the generalized Stokes system. The original formulation of this result in [BP07, Theorem 4.1] allows for different kinds of conditions on different open and closed disjoint components of the compact boundary of the domain $\mathcal{D}$, so that $\mathcal{U}_{A_*}$ and $\mathcal{P}_{A_*}$ are well-defined and they yield the estimates needed.

6.2 The Problem in Two Space Dimensions

Similarly to the first section, we argue that the main preliminary results like maximal regularity of the Stokes and generalized Stokes problem, pressure estimates and the embeddings in Section 1.4 hold for all dimensions $n \geq 2$. The main changes regard the balance equations for the rigid body velocities. They become simpler due to the fact that the set of rigid motions in two dimensions can be described by three parameters instead of six.

The model equations are given by

$$
\begin{cases}
v_t - \operatorname{div} \mathbf{T}(v,q) + (v \cdot \nabla)v &= \mathbf{g} \quad \text{in } J_T \times \mathcal{D}(t), \\
\operatorname{div} v &= 0 \quad \text{in } J_T \times \mathcal{D}(t), \\
v &= v_{\mathcal{B}} \quad \text{on } J_T \times \Gamma(t), \\
v(0) &= v_0 \quad \text{on } \mathcal{D}(0), \\
m\eta' + \int_{\Gamma(t)} \mathbf{T}(v,q)n \, d\sigma &= m\mathbf{g} \quad \text{in } J_T, \\
(J\omega)' + \int_{\Gamma(t)} (x - x_c(t))^{\perp} \cdot \mathbf{T}(v,q)n \, d\sigma &= 0 \quad \text{in } J_T, \\
\eta(0) &= \eta_0, \\
\omega(0) &= \omega_0,
\end{cases}
\tag{6.3}
$$

where we import most of the notation from the previous sections. The angular velocity ω of the body is now $\mathbb{R}$-valued and we write $\begin{pmatrix} x_1 \\ x_2 \end{pmatrix}^{\perp} = \begin{pmatrix} x_2 \\ -x_1 \end{pmatrix}$ for all $(x_1, x_2)^T \in \mathbb{R}^2$. The full velocity of the body is given by

$$
v_{\mathcal{B}}(t,x) = \eta(t) + \omega(t)(x - x_c(t))^{\perp}.
$$

Additionally, we define $J := \int_{\mathcal{B}} \rho_{\mathcal{B}}(x)|x|^2 \, dx$. The following result holds true.

Theorem 6.4. *Let $\mathcal{B}$ and $\mathcal{D}$ be domains of class $C^{2,1}$ in $\mathbb{R}^2$ analogously to the description in Section 2.1. Assume that $1 < p,q < \infty$, $\frac{1}{q} + \frac{1}{p} \leq \frac{3}{2}$, $\frac{1}{2q} + \frac{1}{p} \neq 1$, $\eta_0 \in \mathbb{R}^2$, $\omega_0 \in \mathbb{R}$ and that $v_0 \in B_{q,p}^{2-2/p}(\mathcal{D})$ satisfies the compatibility conditions $\operatorname{div} v_0 = 0$ and $v_0|_{\Gamma}(x) = \eta_0 + \omega_0 x^{\perp}$, if $\frac{1}{2q} + \frac{1}{p} < 1$ and $(v_0|_{\Gamma} \cdot n)(x) = (\eta_0 + \omega_0 x^{\perp}) \cdot n(x)$, $x \in \Gamma$, if $\frac{1}{2q} + \frac{1}{p} > 1$. Then there exists a maximal interval J_T, $T > 0$, such that problem (6.3) admits a unique strong solution*

$$
\begin{aligned}
v &\in L^p(J_T; W^{2,q}(\mathcal{D}(\cdot))) \cap W^{1,p}(J_T; L^q(\mathcal{D}(\cdot))), \\
q &= q_0 + \mathbf{g} \cdot Y, \quad q_0 \in L^p(J_T; \widehat{W}^{1,q}(\mathcal{D}(\cdot))), Y \in C^1(J_T; C^{\infty}(\mathcal{D}(\cdot))), \\
\eta &\in W^{1,p}(J_T), \\
\omega &\in W^{1,p}(J_T).
\end{aligned}
$$

Remark 6.5. Theorem 6.4 can be modified as in Section 6.1 to hold in the case when fluid and body are contained in a bounded domain $\mathcal{O} \subset \mathbb{R}^2$ of class $C^{2,1}$. In this case, the condition $v|_{\partial\mathcal{O}} = 0$ must be included in (6.3).

Again, we do not give a proper proof but explain how the arguments in the previous chapters can be adapted to this situation. The definition and the estimates for the change of coordinates are straightforward. It is also clear how to derive the transformed system of equations in $J_T \times \mathcal{D}$ and its linearization. The reformulation of the linearized system in Section 3.1 depends more strongly on the geometric properties of the body. We have to make the following changes to describe the added mass effect. Instead of the correction v in (3.7) we consider $\overline{v} := \xi_1 \overline{v}^{(1)} + \xi_2 \overline{v}^{(2)} + \Omega \overline{V}$, where ξ is the transformed translational and Ω is the transformed scalar angular velocity of the body and where the $\overline{v}^{(i)}$s and $\overline{V}$ solve

$$
\left\{
\begin{array}{rcll}
\Delta \overline{v}^{(i)} & = & 0 & \text{in } \mathcal{D}, \\
\frac{\partial \overline{v}^{(i)}}{\partial N}\big|_\Gamma & = & N_i & \text{on } \Gamma,
\end{array}
\right.
\quad \text{and} \quad
\left\{
\begin{array}{rcll}
\Delta \overline{V} & = & 0 & \text{in } \mathcal{D}, \\
\frac{\partial \overline{V}}{\partial N}\big|_\Gamma & = & N \cdot y^\perp & \text{on } \Gamma.
\end{array}
\right.
$$

The added mass matrix reduces to

$$
\overline{\mathbb{M}} =
\begin{pmatrix}
\int_\Gamma \overline{v}^{(1)} N_1 \, d\sigma & \int_\Gamma \overline{v}^{(2)} N_1 \, d\sigma & \int_\Gamma \overline{V} N_1 \, d\sigma \\
\int_\Gamma \overline{v}^{(1)} N_2 \, d\sigma & \int_\Gamma \overline{v}^{(2)} N_2 \, d\sigma & \int_\Gamma \overline{V} N_2 \, d\sigma \\
\int_\Gamma \overline{v}^{(1)} y^\perp \cdot N \, d\sigma & \int_\Gamma \overline{v}^{(2)} y^\perp \cdot N \, d\sigma & \int_\Gamma \overline{V} y^\perp \cdot N \, d\sigma
\end{pmatrix}
$$

and it is still positive definite since for all $z = (x_1, x_2, y)^T \in \mathbb{R}^3$,

$$
\begin{aligned}
z^T \overline{\mathbb{M}} z & = \sum_{i=1}^{2} \int_\mathcal{D} \Big[(x_1^2 (\partial_i \overline{v}^{(1)})^2 + x_2^2 (\partial_i \overline{v}^{(2)})^2 + y^2 (\partial_i \overline{V})^2 \\
& \quad + 2 x_1 x_2 (\partial_i \overline{v}^{(1)})(\partial_i \overline{v}^{(2)}) + 2 x_1 y (\partial_i \overline{v}^{(1)})(\partial_i \overline{V}) + 2 x_2 y (\partial_i \overline{v}^{(2)})(\partial_i \overline{V}) \Big] \\
& = \sum_{i=1}^{2} \int_\mathcal{D} \big(x_1 (\partial_i \overline{v}^{(1)}) + x_2 (\partial_i \overline{v}^{(2)}) + y (\partial_i \overline{V}) \big)^2 \\
& \geq 0
\end{aligned}
$$

by the Gauss Theorem. This fact implies that the linear problem in two dimensions can be rewritten as in (3.15) and the corresponding estimates follow as in Section 3.2.

Based on this result, the fixed point argument in Chapter 4 carries over to the two-dimensional case with less requirements on regularity. For $n = 2$, the embeddings needed follow from Proposition 1.8 for all $1 < p, q < \infty$ such that $\frac{1}{p} + \frac{1}{q} \leq \frac{3}{2}$.

Remark 6.6. Again, we obtain an analogous result in the generalized Newtonian case. The estimates and results on the operators $\mathcal{U}_{A_*}$ and $\mathcal{P}_{A_*}$ can be given for any space dimension $n \geq 2$ and the definition and structure of the operators A, A_* and $\mathcal{A}$ does not depend on n in a crucial way. The embedding (5.13) holds true if $p > n + 2$, so that in two dimensions it is sufficient to assume $p > 4$ in order to get a local strong solution.

Bibliography

[AF03]   R. A. Adams and J. J. F. Fournier, *Sobolev Spaces*, Pure and
           Applied Mathematics (Amsterdam), vol. 140, Elsevier/Academic
           Press, Amsterdam, 2003.

[Ama94]   H. Amann, *Stability of the rest state of a viscous incompressible
           fluid*, Arch. Rational Mech. Anal. **126** (1994), no. 3, 231–242.

[Ama95]   ———, *Linear and Quasilinear Parabolic Problems. vol. I*, Mono-
           graphs in Mathematics, vol. 89, Birkhäuser, Boston, 1995.

[Ama00]   ———, *On the strong solvability of the Navier-Stokes equations*,
           J. Math. Fluid Mech. **2** (2000), no. 1, 16–98.

[AS03]   N. Arada and A. Sequeira, *Strong steady solutions for a general-
           ized Oldroyd-B model with shear-dependent viscosity in a bounded
           domain*, Math. Models Methods Appl. Sci. **13** (2003), no. 9, 1303–
           1323.

[AS05]   ———, *Steady flows of shear-dependent Oldroyd-B fluids around
           an obstacle*, J. Math. Fluid Mech. **7** (2005), no. 3, 451–483.

[BMR09]   M. Bulicek, J. Málek, and K. R. Rajagopal, *Mathematical analy-
           sis of unsteady flows of fluids with pressure, shear-rate, and tem-
           perature dependent material moduli that slip at solid boundaries*,
           SIAM J. Math. Anal. **41** (2009), no. 2, 665–707.

[Bog79]   M. E. Bogovskiĭ, *Solution of the first boundary value problem
           for an equation of continuity of an incompressible medium*, Dokl.
           Akad. Nauk SSSR **248** (1979), 1037–1040.

[BP07]   D. Bothe and J. Prüss, L^p-theory for a class of non-Newtonian
           fluids, SIAM J. Math. Anal. **39** (2007), 379–421.

[Che88]   N. P. Cheremisinoff, *Encyclopedia of Fluid Mechanics*, vol. 7, Gulf Publishing, Houston, 1988.

[CSMT00]   C. Conca, J. San Martín, and M. Tucsnak, *Existence of solutions for the equations modelling the motion of a rigid body in a viscous fluid*, Comm. Partial Differential Equations **25** (2000), 1019–1042.

[CT06]   P. Cumsille and M. Tucsnak, *Wellposedness for the Navier-Stokes flow in the exterior of a rotating obstacle*, Math. Methods Appl. Sci. **29** (2006), 595–623.

[CT08]   P. Cumsille and T. Takahashi, *Wellposedness for the system modelling the motion of a rigid body of arbitrary form in an incompressible viscous fluid*, Czechoslovak Math. J. **58 (133)** (2008), 961–992.

[DE99]   B. Desjardins and M. J. Esteban, *Existence of weak solutions for the motion of rigid bodies in a viscous fluid*, Arch. Ration. Mech. Anal. **146** (1999), 59–71.

[DGH09]   E. Dintelmann, M. Geißert, and M. Hieber, *Strong L^p-solutions to the Navier-Stokes flow past moving obstacles: the case of several obstacles and time dependent velocity*, Trans. Amer. Math. Soc. **361** (2009), 653–669.

[DHP07]   R. Denk, M. Hieber, and J. Prüss, *Optimal L^p-L^q-estimates for parabolic boundary value problems with inhomogeneous data*, Math. Z. **257** (2007), 193–224.

[Din07]   E. Dintelmann, *Fluids in the Exterior Domain of Several Moving Obstacles*, Ph.D. thesis, Technische Universität Darmstadt, 2007.

[DR05]   L. Diening and M. Ružička, *Strong solutions for generalized Newtonian fluids*, J. Math. Fluid Mech. **7** (2005), 413–450.

[Fei03]   E. Feireisl, *On the motion of rigid bodies in a viscous compressible fluid*, Arch. Ration. Mech. Anal. **167** (2003), 281–308.

[FHN08]   E. Feireisl, M. Hillairet, and Š. Nečasová, *On the motion of several rigid bodies in an incompressible non-Newtonian fluid*, Nonlinearity **21** (2008), 1349–1366.

[FM70]   H. Fujita and H. Morimoto, *On fractional powers of the Stokes operator*, Proc. Japan Acad. **46** (1970), no. 10, 1141–1143.

[FMS03]   J. Frehse, J. Málek, and M. Steinhauer, *On analysis of steady flows of fluids with shear-dependent viscosity based on the Lipschitz truncation method*, SIAM J. Math. Anal. **34** (2003), no. 5, 1064–1083.

[FNN07]   R. Farwig, Š. Nečasová, and J. Neustupa, *On the essential spectrum of a Stokes-type operator arising from flow around a rotating body in the L^q-framework*, Kyoto Conference on the Navier-Stokes Equations and their Applications, RIMS Kôkyûroku Bessatsu, B1, Res. Inst. Math. Sci. (RIMS), Kyoto, 2007, pp. 93–105.

[FR08]   J. Frehse and M. Ružička, *Non-homogeneous generalized Newtonian fluids*, Math. Z. **260** (2008), no. 2, 355–375.

[Gal94]   G. P. Galdi, *An Introduction to the Mathematical Theory of the Navier-Stokes Equations. Vol. I*, Springer, New York, 1994.

[Gal99]   _____, *On the steady self-propelled motion of a body in a viscous incompressible fluid*, Arch. Ration. Mech. Anal. **148** (1999), no. 1, 53–88.

[Gal02]   _____, *On the motion of a rigid body in a viscous liquid: a mathematical analysis with applications*, Handbook of mathematical fluid dynamics, Vol. I (S. J. Friedlander and D. Serre, eds.), North-Holland, Amsterdam, 2002, pp. 653–791.

[Gal06]   _____, *Determining modes, nodes and volume elements for stationary solutions of the Navier-Stokes problem past a three-dimensional body*, Arch. Ration. Mech. Anal. **180** (2006), no. 1, 97–126.

[Gal07]   _____, *Further properties of steady-state solutions to the Navier-Stokes problem past a three-dimensional obstacle*, J. Math. Phys. **48** (2007), no. 6, 065207, 43.

[GHH06a] M. Geißert, H. Heck, and M. Hieber, *L^p-theory of the Navier-Stokes flow in the exterior of a moving or rotating obstacle*, J. Reine Angew. Math. **596** (2006), 45–62.

[GHH06b] _____, *On the equation $\mathrm{div}\, u = g$ and Bogovskiĭ's operator in Sobolev spaces of negative order*, Partial differential equations and functional analysis, Oper. Theory Adv. Appl., vol. 168, Birkhäuser, Basel, 2006, pp. 113–121.

[GHHS]   M. Geißert, H. Heck, M. Hieber, and O. Sawada, *Weak Neumann implies Stokes*, in preperation.

[GHS97]   G. P. Galdi, J. G. Heywood, and Y. Shibata, *On the global existence and convergence to steady state of Navier-Stokes flow past an obstacle that is started from rest*, Arch. Rational Mech. Anal. **138** (1997), no. 4, 307–318.

[GLS00]   M. D. Gunzburger, H.-C. Lee, and G. A. Seregin, *Global existence of weak solutions for viscous incompressible flows around a moving rigid body in three dimensions*, J. Math. Fluid Mech. **2** (2000), 219–266.

[GS02]   G. P. Galdi and A. L. Silvestre, *Strong solutions to the problem of motion of a rigid body in a Navier-Stokes liquid under the action of prescribed forces and torques*, Nonlinear Problems in Mathematical Physics and Related Topics I, Kluwer Academic/Plenum Publishers, New York, 2002, pp. 121–144.

[GS05]   ———, *Strong solutions to the Navier-Stokes equations around a rotating obstacle*, Arch. Ration. Mech. Anal. (2005), no. 3, 331–350.

[GS07]   ———, *The steady motion of a Navier-Stokes liquid around a rigid body*, Arch. Ration. Mech. Anal. **184** (2007), no. 3, 371–400.

[GV01]   G. P. Galdi and A. Vaidya, *Translational steady fall of symmetric bodies in a Navier-Stokes liquid, with application to particle sedimentation*, J. Math. Fluid Mech. **3** (2001), no. 2, 183–211.

[GVP+02] G. P. Galdi, A. Vaidya, M. Pokorný, D. D. Joseph, and J. Feng, *Orientation of symmetric bodies falling in a second-order liquid at nonzero Reynolds number*, Math. Models Methods Appl. Sci. **12** (2002), no. 11, 1653–1690.

[Har64]   P. Hartman, *Ordinary Differential Equations*, John Wiley and Sons, New York, 1964.

[His99]   T. Hishida, *An existence theorem for the Navier-Stokes flow in the exterior of a rotating obstacle*, Arch. Ration. Mech. Anal. **150** (1999), 307–348.

[Hop51]   E. Hopf, *Über die Anfangswertaufgabe für die hydrodynamischen Grundgleichungen*, Math. Nachr. **4** (1951), 213–231.

[HS99]   K.-H. Hoffmann and V. Starovoitov, *On a motion of a solid body in a viscous fluid. Two-dimensional case*, Adv. Math. Sci. Appl. **9** (1999), 633–648.

[HS09]   T. Hishida and Y. Shibata, $L_p - L_q$ *estimate of the Stokes operator and Navier-Stokes flows in the exterior of a rotating obstacle*, Arch. Rational Mech. Anal. **193** (2009), no. 2, 339–421.

[HT09]   M. Hillairet and T. Takahashi, *Collisions in three-dimensional fluid structure interaction problems*, SIAM J. Math. Anal. **40** (2009), no. 6, 2451–2477.

[IW77]   A. Inoue and M. Wakimoto, *On existence of solutions of the Navier-Stokes equation in a time dependent domain*, J. Fac. Sci. Univ. Tokyo Sect. IA Math. **24** (1977), no. 2, 303–319.

[Jos90]   D. D. Joseph, *Fluid Dynamics of Viscoelastic Liquids*, Applied Mathematical Sciences, vol. 84, Springer, New York, 1990.

[Kir69]   G. Kirchhoff, *Über die Bewegung eines Rotationskörpers in einer Flüssigkeit*, J. Reine Angew. Math. **71** (1869), 237–281.

[Lad69]   O. A. Ladyzhenskaya, *The mathematical theory of viscous incompressible flow*, Mathematics and its Applications, vol. 2, Gordon and Breach, New York, 1969.

[LBS07]   E. Lauga, M. P. Brenner, and H. A. Stone, *Microfluidics: The no-slip boundary condition*, Handbook of Experimental Fluid Dynamics (J. Foss, C. Tropea, and Y. Yarin, eds.), Springer, New York, 2007, pp. 1219–1240.

[Ler34]   J. Leray, *Sur le mouvement d'un liquide visqueux emplissant l'espace*, Acta Math. **63** (1934), 193–248.

[LM00]   P. L. Lions and N. Masmoudi, *Global solutions for some Oldroyd models of non-Newtonian flows*, Chinese Ann. Math. Ser. B **21** (2000), no. 2, 131–146.

[MNR93]   J. Málek, J. Nečas, and M. Ružička, *On the non-Newtonian incompressible fluids*, Math. Models Methods Appl. Sci. **3** (1993), 35–63.

[MNR01]   ———, *On weak solutions to a class of non-Newtonian incompressible fluids in bounded domains: The case $p \geq 2$*, Adv. Differential Equations **6** (2001), 257–302.

[MR05]   J. Málek and K. R. Rajagopal, *Mathematical issues concerning the Navier-Stokes equations and some of their generalizations*, Handbook of Differential Equations, Evolutionary Equations (C. Dafermos and E. Feireisl, eds.), vol. 2, Elsevier, New York, 2005, pp. 371–459.

[MRR95]   J. Málek, K. R. Rajagopal, and M. Ružička, *Existence and regularity of solutions and the stability of the rest state for fluids with shear dependent viscosity*, Math. Models Methods Appl. Sci. **5** (1995), no. 6, 789–812.

[NS03]   A. Noll and J. Saal, *H^∞-calculus for the Stokes operator on L^q-spaces*, Math. Z. **244** (2003), 651–688.

[NSV99]   A. Novotny, A. Sequeira, and J. Videman, *Steady motions of viscoelastic fluids in three-dimensional exterior domains. Existence, uniqueness and asymptotic behaviour*, Arch. Ration. Mech. Anal. **149** (1999), no. 1, 49–67.

[Ren00]   M. Renardy, *Mathematical Analysis of Viscoelastic Flows*, CBMS-NSF Regional Conference Series in Applied Mathematics, vol. 73, SIAM, Philadelphia, 2000.

[Sal05]   R. Salvi, *Existence and uniqueness results for non-Newtonian fluids of the Oldroyd type in unbounded domains*, Regularity and Other Aspects of the Navier-Stokes Equations, Banach Center Publ., vol. 70, Polish Acad. Sci., Warsaw, 2005, pp. 209–237.

[Sau72]   N. Sauer, *The steady state Navier-Stokes equations for incompressible flows with rotating boundary*, Proc. Roy. Soc. Edinburgh A **110** (1972), 93–99.

[Ser87]   D. Serre, *Chute libre d'un solide dans un fluide visqueux incompressible. Existence*, Japan J. Appl. Math. **4** (1987), 99–110.

[SMST02]   J. San Martín, V. Starovoitov, and M. Tucsnak, *Global weak solutions for the two-dimensional motion of several rigid bodies in an incompressible viscous fluid*, Arch. Ration. Mech. Anal. **161** (2002), 113–147.

[Soh01]   H. Sohr, *The Navier-Stokes equations. An Elementary Functional Analytic Approach*, Birkhäuser, Basel, 2001.

[Sol77] V. A. Solonnikov, *Estimates for solutions of nonstationary Navier-Stokes equations*, J. Sov. Math. **8** (1977), 467–529.

[SS07] Y. Shibata and R. Shimida, *On a generalized resolvent estimate for the Stokes system with Robin boundary condition*, J. Math. Soc. Japan **59** (2007), no. 2, 469–519.

[Sto51] G. Stokes, *On the effect of internal friction of fluids on the motion of pendulums*, Trans. Cambridge Phil. Soc. **9** (1851), 8–85.

[Tak03] T. Takahashi, *Analysis of strong solutions for the equations modeling the motion of a rigid-fluid system in a bounded domain*, Adv. Differential Equations **8** (2003), no. 12, 1499–1532.

[Tem79] R. Temam, *Navier-Stokes Equations*, North Holland, Amsterdam, 1979.

[Tri95] H. Triebel, *Interpolation Theory, Function Spaces, Differential Operators*, Johann Ambrosius Barth, Heidelberg, 1995.

[TT04] T. Takahashi and M. Tucsnak, *Global strong solutions for the two-dimensional motion of an infinite cylinder in a viscous fluid*, J. Math. Fluid Mech. **6** (2004), 63–77.

[Wei73] H.F. Weinberger, *On the steady fall of a body in a Navier-Stokes fluid*, Proc. Symp. Pure Math. **23** (1973), 421–440.

[Wie05] K. Wielage, *Analysis of Non-Newtonian Two-Phase Flow*, Ph.D. thesis, University of Paderborn, 2005.